HARVARD | BUSINESS | SCHOOL

EXECUTIVE EDUCATION

High Potentials Leadership Program

Offered by the

Harvard Business School Leadership Initiative

Anthony J Mayo

Entrepreneurs, Managers, and Leaders

Entrepreneurs, Managers, and Leaders

What the Airline Industry Can Teach Us about Leadership

Anthony J. Mayo,
Nitin Nohria, and
Mark Rennella

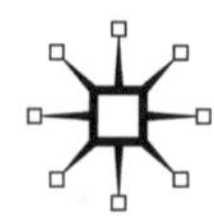

ENTREPRENEURS, MANAGERS, AND LEADERS

First published in 2009 by
PALGRAVE MACMILLAN®
in the United States—a division of St. Martin's Press LLC,
175 Fifth Avenue, New York, NY 10010.

Where this book is distributed in the UK, Europe and the rest of the world, this is by Palgrave Macmillan, a division of Macmillan Publishers Limited, registered in England, company number 785998, of Houndmills, Basingstoke, Hampshire RG21 6XS.

Palgrave Macmillan is the global academic imprint of the above companies and has companies and representatives throughout the world.

Palgrave® and Macmillan® are registered trademarks in the United States, the United Kingdom, Europe and other countries.

ISBN: 978–0–230–61567–0

Library of Congress Cataloging-in-Publication Data

Mayo, Anthony J.
Entrepreneurs, managers, and leaders : what the airline industry can teach us about leadership / Anthony J. Mayo, Nitin Nohria, Mark Rennella.
p. cm.
Includes bibliographical references and index.
ISBN 978–0–230–61567–0
1. Airlines—United States—Management. 2. Leadership. I. Nohria, Nitin, 1962– II. Rennella, Mark. III. Title.

HE9803.A4M39 2009
658.4'092—dc22 2008053005

A catalogue record of the book is available from the British Library.

Design by Newgen Imaging Systems (P) Ltd., Chennai, India.

First edition: September 2009

10 9 8 7 6 5 4 3 2 1

Printed in the United States of America.

To our children

Hannah, Alexander, and Jacob Mayo
Reva and Ambika Nohria
Davis and Benjamin Rennella

Contents

Acknowledgments

The rise, bitter fall, and rebirth of the airline industry over the past one-hundred years captured our imaginations and sparked our interest in exploring industry evolution and specifically the role that leadership played in shaping that evolution. In many ways, the research and writing of *Entrepreneurs, Managers, and Leaders* was, for us, an evolutionary process—one that took us down many different avenues. Along the way, we received a considerable amount of support for which we are very grateful.

We are thankful for the support and encouragement provided by Harvard Business School, especially Dean Jay Light and the Faculty Chairs of the Division of Research, Professors Debora Spar and Srikant Datar. We also wish to acknowledge the support of other members of the Division of Research, including Research Director, Professor Geoff Jones, Ann Cichon, and Steve O'Donnell. In particular, we would like to acknowledge Linda Hill and the HBS Leadership Initiative for supporting us throughout this project. We are indebted to the staff of Baker Library, especially Laura Linard and Melissa Murphy of the Historical Collections Department, who helped to identify and provide archival photos and information about the early years of aviation in the United States.

Throughout the process of researching and writing *Entrepreneurs, Managers, and Leaders* we have benefited from the insights and fresh perspectives of a number of friends and colleagues, including Tom Gaffny, Rakesh Khurana, Joshua Margolis, and Scott Snook. In particular, we would like to acknowledge Christopher Marquis who read an early draft of the entire manuscript and provided thoughtful and actionable feedback. We are grateful for Laura Singleton's assistance on the C. R. Smith chapter, Mark Benson's work on the Gordon Bethune chapter, and Michael Horn's insights and background on the early years of Juan Trippe. At various times throughout the last few years, Deborah Bell, Letty Garcia, Amanda Pepper, and Lisa Pode provided excellent research assistance and administrative support. Amanda Pepper was very helpful in conducting the early rounds of photo research for the book. Finally, we wish to thank our publisher at Palgrave Macmillan, Laurie Harting, who provided insightful advice and suggestions during the final editing process.

I (Tony Mayo) wish to thank my co-authors Nitin and Mark for their support and dedication to this project. *Entrepreneurs, Managers, and Leaders* is the culmination of a truly collaborative process in which we explored a variety of conceptual models and a vast array of literature. I am deeply indebted to Nitin and Mark for their commitment and enthusiasm throughout this journey. Each of us is fascinated with aviation and that sense of wonder and awe enabled us to work together productively and have fun in the process. I have greatly enjoyed sharing this journey with my family. My wife, Denise, and our three children, Hannah, Alexander, and Jacob, have been a tremendous source of inspiration and encouragement for me. I wish to thank them for their steadfast love and support.

Some of my (Nitin Nohria's) best memories are travels on airplanes—from my boarding school to my home in Calcutta when I was just five years old, flying on newly introduced Boeing 707s under the care of gracious air hostesses, to traveling when I had just become an adult on a Boeing 747 (a Jumbo jet as it was affectionately called) to leave India for America, which eventually became my new home country. Now, my children complain that I live on an airplane, given the amount of travel I routinely do (I am a member of the million mile club on three different airlines). Understanding how an industry that does so much to make the world a smaller place can yet be one that has lost so much money was one of the conundrums that drew me to this exploration of the evolution of the airline industry. And what a journey it has been. Working with Tony and Mark, the best fellow travelers I could have hoped for, has led us to better understand the ups and downs of the airline industry and the role that different types of executives have played in different stages in its evolution. It has also led us to think more broadly about how leaders can influence the evolution of any industry—for better or for worse.

I (Mark Rennella) am very grateful to have worked on this project with Tony and Nitin. In addition to their expertise and experience in business, Tony and Nitin both have a talent of bringing out the best in the people they work with. Projects this enjoyable and interesting do not come around every day and I was very fortunate to be invited to contribute to this one. In my years working on historical projects, I have seen not only how the past helps to shape the present, but that history is the outcome of human *choices*, both large and small. Looking at the past, surprisingly, shows how the present is ripe with possibility. My children—Davis and Ben—show me every day how vibrant that promise is. My deepest thanks, also, to my parents who have believed in my promise from day 1. More to come, Mom and Dad.

Introduction

Since the end of the twentieth century and into the twenty-first century there has been increasing anxiety about celebrating larger than life heroes of business. A considerable amount of the backlash has been brought on by business leaders themselves. We have suffered no shortage of high profile scandals. Leadership in America and especially in Corporate America is severely under question. Not since the great stock market crash of 1929 have business leaders been so vilified. In the 1990s, it was all too easy to fall under the "cult of the CEO"—worshipping business mavericks that seemed to create value from almost nothing. In a recent popular Conventional Wisdom section of *Newsweek* magazine, CEOs were described as follows: "Old [CEO]—Superhero role models who drive the economy. New [CEO]: Greedy chiselers who hurt the economy."[1]

In the past decade, books like *Good to Great* by Jim Collins have trumpeted humility as one of the key personal characteristics of truly great business leaders.[2] The success of *Good to Great* and other books that have celebrated the quiet, thoughtful leader is no surprise given the context of our times. Society is sick of the astronomical pay for CEOs and the lack of correlation between CEO pay and corporate performance.[3]

Although we do not deny that the "cult of the CEO" has run its course, we believe that there is danger in not acknowledging and recognizing the role that leaders have played and will play in shaping businesses and industries. At certain stages in the evolution of a business or an industry, ambition, bravado, and even some hubris are necessary. There is no doubt that certain leaders, who exhibit these traits, have contributed to some cataclysmic downfalls, but they have also created some incredible companies that have profoundly influenced the way we live and work.

In *Entrepreneurs, Managers, and Leaders*, we chronicle the role that leaders played in shaping the evolution of the airline industry—both its ups and downs. The ability to succeed in an industry that is inherently fraught with risk, intense competition, and uncertain geopolitical forces requires a certain type of leadership to push forward. As the airline industry evolved so too did the people who led it. Many of the early leaders were "larger than life" heroes who helped to forge the foundation of an industry. It is certainly conceivable that the uncertain nature of the early aviation industry attracted these types of individuals. During the later stages of the airline industry's evolution, a different breed of leader emerged who shepherded the industry through its growth and maturity. More recently, many airlines are looking for a new type of leader to weather the current and severe economic downturn in the industry.

We tell all three stories in this book—not to resurrect the cult of the great man, but to appropriately highlight the role that leaders play in creating, growing, and reinventing companies and industries. Perhaps as much as any other industry, airlines have historically attracted strong and often charismatic personalities who have shaped and were shaped by their industry. The first generation included entrepreneurs Juan Trippe of Pan American World Airways and Eddie Rickenbacker of Eastern Airlines in the 1920s and 1930s. The second generation was marked by the management expertise of C. R. Smith of American Airlines and Pat Patterson from United Air Lines who developed and reinforced a model for success within the regulated environment of the industry for 40 years. Finally, notable change agents shaped the industry after deregulation in the last decades of the twentieth century such as Herb Kelleher of Southwest Airlines and Gordon Bethune of Continental Airlines.

Despite the indelible mark these individuals made on their industry, we found that writers on industry evolution—concerning the airlines or any other industry—have rarely factored in leadership as a way of explaining or understanding that evolution. Historically, the study of industry evolution has focused less on the role of leaders and more on the technological and environmental factors that defined the competitive landscape and influenced an industry's evolution.[4] Through our analysis of the airline industry in the twentieth century, we seek to paint a fuller picture of the interdependent relationship between the actions of leaders, the context of their times, and the evolution of an industry.

In our study of the history of business in the United States (*In Their Time: The Greatest Business Leaders of the 20th Century*), we discovered that there is a recursive relationship between the actions of business executives and the contextual landscape in which they operate; each influences and shapes the other. The environmental factors that we highlighted—demographic shifts, technological breakthroughs, government regulations, geopolitics, labor conditions, and social mores—coalesce to create a contextual framework for business, within which some individuals envision new enterprises, others see opportunities for greatly expanding the scale and scope of existing businesses, and still others find opportunity through the reinvention or recreation of companies or technologies that were considered stagnant or declining. We called these individuals Entrepreneurs, Managers, and Leaders respectively.[5]

Entrepreneurs, managers, and leaders can be both a product of their context, but they can also shape it in fundamental ways. At different points in time, certain archetypes seem to play a larger role in shaping the context or seem to be poised to capitalize on the prevailing social and economic environment. In the early decades of the twentieth century, Joseph A. Schumpeter examined individuals who were able to build new enterprises through a process of creative destruction and rebuilding. He noted that "We have seen the function of entrepreneurs is to reform or revolutionize the pattern of production by exploiting an invention, or more generally, an untried technological possibility for producing a new commodity or producing an old one in a new way, by opening up a new source of supply of materials or a new outlet for products, by reorganizing an industry and so on."[6] Our choice of the name *entrepreneurs* for individuals who created something new within their context was based on Schumpeter's work.[7]

Looking at executives in large, decentralized organizations in the 1920s and 1930s, Alfred D. Chandler, Jr., chronicled the way certain individuals were able to operate their enterprises in an efficient, expedient, and cost effective manner. He wrote: "Top managers, in addition to evaluating and coordinating the work of middle managers, took the place of the market in allocating resources for future production and distribution. In order to carry out these functions, the managers had to invent new practices and procedures which in time became standard operating methods in managing American production and distribution."[8] In essence, to capitalize on the growth of businesses and industries *managers of managers* were required. For Chandler, these managers played a vital role in shaping market forces.

Later in the twentieth century, Warren Bennis and Burt Nanus studied individuals who were especially adept at reframing or reinventing businesses for sustained success. They wrote: "Effective leadership can move organizations from current to future states, create visions of potential opportunities for organizations, instill within employees commitment to change and instill new cultures and strategies in organizations that mobilize and focus energy and resources . . . They emerge when organizations face new problems and complexities that cannot be solved by unguided evolution. They assume responsibilities for reshaping organizational practices to adapt to environmental changes. They direct organizational changes that build confidence and empower employees to seek new ways of doing things."[9] Leaders are skilled at reconceptualizing a business and galvanizing followers around their vision.

Although Schumpeter, Chandler, Bennis, and other scholars have examined specific leadership archetypes at certain points in time, our research has the advantage of a pan-century viewpoint. By looking at an industry over the course of 100 years, we have been able to witness the interplay of the leadership archetypes and their existence throughout the twentieth century. Through this process, we discovered that our archetypes of Entrepreneur, Manager, and Leader fit nicely into the evolving stages in the lifecycle of industries and businesses (see figure I.1).[10] Although certain points in an industry's lifecycle favor a dominant leadership archetype, we also found in our study of 1,000 business executives across more than 20 industries that contextual landscapes, though unique and evolving, provide opportunities for all 3 leadership archetypes at any point in time.[11] As we delved deeper into the airline industry, we were particularly intrigued by the type of leader that emerged at each stage of the industry's evolution and the role the leader played in influencing the direction of that evolution. We sought to understand how entrepreneurial activity emerged during periods of uncertainty and in contrast what constituted success in periods of relative stability or government oversight. Was the leader a product of his or her times or did he or she create the opportunities for success?

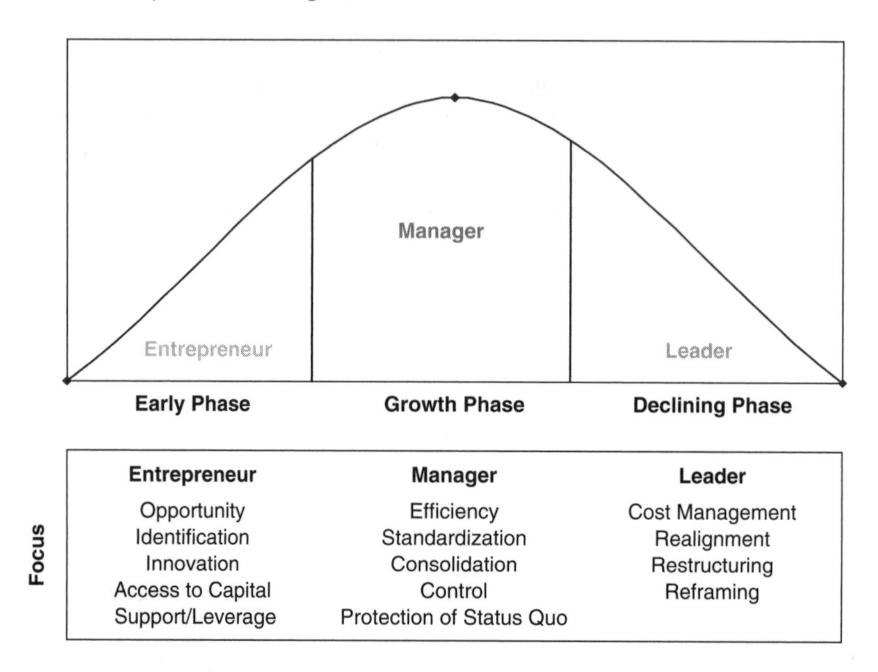

Figure I.1 Archetypal executives across an industry lifecycle

In our study, we found that during periods of great uncertainty, such as the initial emergence of an industry, the actions of individuals can have a large impact on harnessing the evolving contextual landscape. In many cases, the industry becomes a mirror image of the entrepreneur. It should not be surprising then that entrepreneurs play one of the largest roles in shaping the early nature and evolution of an industry. With no clear precedents to guide them and wide-open opportunities, entrepreneurs pursue a trial-and-error approach to establish a business model that has sustainable success. During these periods, it is hard for entrepreneurs to be too excessive or too expansive; having a certain level of hubris and determination is often a perquisite for success during periods of intense change and uncertainty as an industry begins to take shape.

Though hubris can precipitate a downfall if left unchecked, it can be useful for entrepreneurs trying to break into a new industry or forge some new ground. In their study of entrepreneurship, Mathew Hayward, Dean Shepherd, and Dale Griffin note: "Greater overconfidence provides founders with the bravado to undertake and persist with more challenging tasks and the conviction that they will have the necessary resources for their ventures to succeed."[12] On the downside, the authors note: "overconfident founders may exaggerate the utility of their unique personality and leadership skills, relative to competing founders," and in so doing, they may not prepare adequately for anticipated needs.[13] One airline executive admitted as much when he said, "if you're an entrepreneur, you're optimistic by nature. So you think, in six months, we're going to be sailing. But the optimism causes you to raise a lot less capital than you need in most cases, and it's very lonely."[14]

The goal of entrepreneurs is to carve out an approach that can become a significant business model as the industry matures. A variety of business models that reflect the personalities, quirks, and competing visions of different entrepreneurs vie for survival in the early phase of an industry.[15] Eventually, the jockeying for a dominant position settles down and a standard operating model for success is defined. The model for success can be defined by the market through customer preferences, operational efficiency, or economies of scale. The model can also be reinforced through external factors such as demographic shifts or government regulation.[16] When a dominant business model emerges, it sets the stage for the emergence of the next leadership archetype.

At this stage, the entrepreneur archetype gives way to the manager archetype. Managers focus on growing, elaborating, refining, and protecting this dominant design—greatly expanding its scale and scope.[17] The experimental frenzy that characterizes the start-up phase is typically replaced by stability and structure. Managers are far from reactive during these times. They often seek ways to expand product features or services to solidify their dominance. In essence, managers act as supporters or promoters of a contextual framework that reinforces the dominant business model. Managers during this period turn their attention to standardization, efficiency, and consolidation. Managers also seek opportunities to make their products or services more attractive and appealing to a broader customer base. In some cases, however, the earnest efforts of managers to protect their dominant business models actually create the seeds of their own destruction. So much effort and attention is paid to the status quo that there is often a failure to focus on the evolving competitive landscape and the changing environmental conditions.[18]

Savvy entrepreneurs and change agents often create niche opportunities to exploit the complacency and stasis that constricts change for established players. Threatened by new entrants or diverse business models, entrenched managers often do more of the same—investing in a model that may no longer be relevant. Managers may be unable to respond quickly to changes or may be too entrenched in bureaucracies to recognize the need for change.[19] The airline industry under regulation was particularly susceptible to this potential flaw. In some cases, it created a false sense of security, further reducing the need or desire for change.

The conditions that coalesce to challenge the formerly dominant business model including saturation of prevailing demand, diminished ability to raise prices combined with a high cost structure and overcapacity, rigidities in labor and other supply contracts, and loss of public support for protective regulations become the fertile playground for *leaders* or change agents who are skilled at reinvention.[20] Taking charge of a business in a state of decline, leaders seek opportunities to recreate or transform a dominant business model to renew them to their former stature in the industry. During these intense inflection periods within industries, leaders emerge to make sense of the chaos and define a new business model that is more aligned with the changing contextual landscape. This model may have the potential to become the new dominant business model, or alternatively, it may set the stage for a parallel opportunity for success. Leaders reintroduce variation and change into the stability of the past to create new opportunities for success, and in so doing they help to regenerate the lifecycle of the entire industry.

Through our analysis of the airline industry, we hope to shed light on the interplay between contextual and individual forces in industry evolution and to specifically explore the role entrepreneurs, managers, and leaders play at different stages of an industry's lifecycle. While every industry has its unique characteristics, all industries are shaped by a similar co-evolution of contextual factors and leadership forces.

The most successful people, like the most successful companies, evolve through time. And although the results of this change are often fruitful, we all know that accepting the need to change and *then* enacting changes—whether in an individual's life or in the evolution of a large corporation—can be extremely difficult. *In Their Time*, the sister volume of this study of CEOs in the airline industry, shows how complicated it is for a CEO to navigate any corporation toward success. He or she must not only ensure that the company's performance measures up against the challenge of its competitors in the here and now, but must also clearly assess the future problems and opportunities stemming from various factors outside of the company—such as evolving technologies or the forces of globalization—over which a CEO has, at best, limited control.[21]

Imagine the hurdle faced by a CEO that has successfully brought his company through one phase of its growth and then is confronted with new business challenges in a context that has greatly changed since his ascension to the front office. If, for instance, an entrepreneurial CEO grows her company to the point that its future development is now dependent on good management skills, *can this CEO change, too*? Can she transform not only how her company functions, but how she personally functions as well?

As many CEOs of failed businesses have learned, the complex landscape of industry evolution often only becomes clear in hindsight. Through telling the story of an industry from the point of view of the CEOs who helped shape it (and were in turn shaped by it), we hope to offer readers a birds-eye view of the way individuals navigate and impact a dynamic and evolving industry landscape. As we will see, the idiosyncrasies of the leaders who were attracted to the industry in its infancy led to the creation of very different kinds of companies. Furthermore, the cumulative temperament of a handful of powerful managers within the airline industry was decisive in shaping a dominant business model for many decades. Finally, as the industry has matured and fallen into very difficult times, a few leaders have risen to the challenge to provide new ideas for the way forward.

Airline Industry Evolution

Why airline CEOs? The story of the airline industry since its inception in the early years of the twentieth century, as much as any business in America, is marked by dramatic changes in the context in which CEOs had the opportunity to forge their identity and the fortunes of their companies. Early on in their history, airlines had to adapt to rapid changes in the technology of flight while attempting to assuage the fears of consumers who believed that planes were inherently dangerous. Later on in their history, labor problems punctuated the evolution of the airline industry in which striking pilots and mechanics often shut down companies completely. U.S. airlines began as potent forces of globalization expanding the reach of U.S. power, but would later struggle to maintain their dominance in a global market that was populated by scores of foreign competitors. After having been one of the most heavily regulated industries for 40 years, airlines were suddenly fully deregulated in 1978. These extraordinary shifts in the external context make the airline industry a fascinating backdrop against which we can examine the role of individual leaders. If we can see the influence of leaders in an environment in which contextual forces are so powerful, surely the importance of individual leaders in the evolution of other industries deserves more careful attention.[22]

The airline industry in the United States over the course of the twentieth century experienced both long periods of stability and revolutionary inflection periods (see table I.1). The periods of stability were influenced by a dominant business model that was reinforced by the contextual environment (specifically government regulation) and the actions of the industry's primary business managers who worked to protect this dominance. The inflection periods were created through a "shakeout" of the industry as it progressed from start-up to stability and through more systematic environmental factors, which contributed to the move toward deregulation.[23] We will explore each phase of the airline industry's evolution through the stories of CEOs who influenced and were influenced by that phase.

Phase I—Start-Up

Part I of this book will capture the different versions of entrepreneurship that were pursued in the early days of the airline industry. In the absence of a dominant business model for success, there was ample opportunity for experimentation. Although the Wright Brothers' achievements in developing airplanes began in 1903, many basic elements in the industry were unresolved well into the 1920s. The personalities that entered the industry and the business models they pursued varied greatly. A key contextual factor, the broad acceptance of flight by the American public, remained in flux for a very long time, thanks, in large part, to the influence of daredevil "birdmen" and "barnstormers" in this early period, whose individualistic vision of how to benefit from the industry competed with those who were interested in building more expansive businesses. Equally importantly, advances in aircraft technology were hampered by legal disputes between manufacturing companies, most of which were ironically instigated by the Wrights themselves.

The stubbornness with which the Wrights tried to protect the technological evolution of their invention greatly influenced the early dynamics of the U.S. airline industry. In America, there were keen intranational rivalries between manufacturers, such as the Wrights and Glenn Curtiss. These disputes

Table I.1 Evolution of the airline industry in the United States

	Start-Up (1903–1930)	*Inflection Period (1930–1938)*	*Growth (1938–1968)*	*Inflection Period (1968–1978)*	*Maturity/Decline (1978–2001)*	*Inflection Period (2001–)*
Major Shifts		Consolidation Scale / Scope Technology Regulation		Succession Saturation Costs (Oil)		Competition Costs (Oil) Terrorism
Contextual Factors	– Experimentation – Competition from railroads & ships – European airline industry competition – Charles Lindbergh's flight in 1927		– Government regulation – Consumer fear / safety – World War II impact on technology and safety – Focus on business traveler		– Deregulation – Massive competition – Saturated market – Lower price points – Broader consumer focus	
Business Model	– Postal subsidies – Crop dusting – Wealthy tourists – Undefined		– Government subsidies – City pair routes – Expansion—demand creation – Domestic focus		– Hub & spoke – Point-to-point – International focus – Partnerships – Consumer focus (loyalty programs)	
Technology	– Emerging – Many models, lack of structure and stability		– Standardized model (DC-3) – Jets – Jumbo Jets – Technology creep—different planes for different routes		– Larger Jumbo Jets—only two players (Boeing and Airbus) – No significant change in technology in almost 20 years – Fleet standardization	
Leadership Required	– Political savvy – Access to capital – Entrepreneurship – Risk taking		– Political savvy / lobbying ability – Drive technology & innovation – Build demand & market share – Focus on efficiency & standardization		– Focus on consolidation & rationalization of resources – Financial management – Turnaround – Competitive positioning – Labor management – Increased complexity	

prompted lawsuits; as early as 1910, the Wrights brought Curtiss to court, claiming that he had infringed their patent on airplane control systems.[24] This was just the first of many lawsuits, which ensured that the Wrights and many other manufacturers spent as much time in the courts as they did in their workshops and manufacturing plants. Meanwhile, unencumbered by these patent battles, the European airline industry grew quickly as aircraft technology there advanced more rapidly.

In these early days of aviation, the government's relationship with this new industry was also a work-in-progress. Although the Congress allocated huge sums of money to mass-produce a large fleet of airplanes for use in World War I, the early stage of aircraft development in the United States was not very productive. Airplanes were delicate and complicated instruments. Producing them still required the expert touch of skilled or semiskilled laborers. Some aviation experts, alarmed by Europe's technological lead over the United States in manufacturing airplanes as well as the growth of European-owned airlines in Latin America, strongly advocated placing all U.S. aviation under government control.

In chapter 1, we explore the efforts of Harry Guggenheim to promote commercial flight as a safe and viable business. Guggenheim saw the airline industry as a means to connect the United States to the rest of the world and sought ways to promote its potential for commerce. In contrast to the Wrights who fought for patent protections at every turn, Guggenheim supported the sharing of technical information for the overall advancement of the industry. Guggenheim combined his own expertise as a pilot with his family's fortune to advocate the development of aviation as a private industry through research and development as well as shrewd public relations. The story of Guggenheim contrasted with the Wrights shows how an individual who gets scant mention in the historical annals of the aviation industry may have actually had more influence on the eventual growth of the U.S. airline industry than the Wright brothers, whose invention of the airplane has received so much attention that they are almost seen as the most vital forces in promoting this industry.

Although relatively few American airplanes eventually made their mark on European skies by the end of World War I, the U.S. government had succeeded in creating an airplane surplus in America. This was a boon for some entrepreneurs who took advantage of the low prices for surplus planes to start up their own airlines on a shoestring budget. Juan Trippe who eventually grew Pan American World Airways into the largest international air carrier in the United States for much of the twentieth century got his start by purchasing surplus airplanes. In chapter 2, we explore Trippe's early entrepreneurial activities to gain a foothold in the emerging airline industry.

Financial barriers to entry in the nascent industry were relatively low.[25] Airplanes were not very expensive and the infrastructure often consisted of an open field or a strip of undeveloped coastline. In the start-up phase of the airline industry, there were dozens of companies competing to capitalize on the emerging opportunities, but early aviation entrepreneurs had significant problems in making a profit after getting their businesses off the ground. Constrained by the small size and carrying capacity of early airplanes as well as their relatively slow speeds, there was a great amount of experimentation in making commercial aviation profitable. Many early ventures avoided head-to-head competition with the long-established (and faster) railroads, searching out niche markets where planes had a clear advantage over other modes of transportation. But an uncertain consumer base and worries about the safety of this newfangled technology undercut many of these early efforts.

Stepping in to fill in this economic gap was the federal government. Specifically, the post office became interested in using airplanes to carry the U.S. mail. After the army struggled for a few months carrying the mail, the post office took over in August 1918. Then, about six years later, Congress passed the Kelly Act, which allowed the postmaster general to grant airmail contracts to private commercial airlines. This contextual factor decisively shaped the initial competitive landscape for the airline industry. Pan Am, for example, was built on Trippe's ability to secure airmail contracts to deliver mail from the United States to Latin American countries.

Though the founder of Delta Air Lines was locked out of many of the early airmail contracts, C. E. Woolman tried to create a viable airline business through combining his childhood love of aviation with his experience in agricultural management. When he saw an opportunity to combine these interests, Woolman jumped at it. He pioneered crop dusting in the Mississippi Delta region, taking this innovation as far south as Peru in an effort to find ways to keep his dusters in the air for a few months longer than the North American growing season would allow. This involvement in crop dusting eventually steered Woolman toward providing passenger service in the underserved regions connecting Dallas and Atlanta through Woolman's home in Louisiana. We will explore Woolman's efforts to carve out a niche, focused on carrying passengers rather than mail, in the airline industry in chapter 3.

It was only well after World War I that concerns about safety and manufacturing problems became more salient, and the weeding out of business models began. Charles Lindbergh's solo, nonstop flight across the Atlantic from Long Island to Paris on May 20, 1927, and his subsequent tour of the United States that autumn marked the beginning of a sharp decrease in public skepticism, which occurred at the same time that U.S. airplane manufacturing finally came into its own. The young daring pilot became an international celebrity overnight. The attention Lindbergh brought to the potential of air transportation created a flurry of speculation in the airline business, and irrationally exuberant investors threw money at almost anything that flew.

"The public has been convinced," the *Washington Post* triumphantly surmised at the end of Lindbergh's tour of 48 states, "that commercial flying, in the hands of competent pilots using the best ships, will soon become a factor in the transportation history of the United States."[26] But despite the best efforts of aviation promoters like Harry Guggenheim, who did much to assuage the fears of the public, airplane tragedies rather than safety always seemed to garner the headlines. The deaths of famous celebrities such as Knute Rockne (March 31, 1931) and Will Rogers (August 15, 1935) along with a rash of accidents in the mid-1930s collectively overshadowed the efforts of aviation pioneers to change the general impression that, despite improvements in aircraft design, flying was still very risky.

The inflection period between the first (if Lindbergh's flight in 1927 is seen as key juncture of the first stage) and second stages of the airline industry was long because of the significant changes (in number and degree) that had to come to a head before the future of the industry was no longer clouded by major uncertainties. These changes ranged from public perceptions of flying safety, to geopolitical concerns, to technological leaps in innovation, to changes in government policy concerning who would run the industry (the government or private corporations) and how it would eventually be run (through many small companies or a few large companies). By 1938, many of these uncertainties had finally been resolved.

Phase II—Movement toward a Dominant Business Model

A series of events ushered in the mature phase of the airline industry after including (1) the consolidation of many small airlines into a few major airlines during President Herbert Hoover's administration; (2) the dissolution of airline conglomerates under President Franklin Roosevelt's administration, which freed individual airlines to purchase airplanes from any manufacturer; (3) the introduction of the DC-3, the first commercially viable passenger aircraft; and (4) the establishment of the Civil Aeronautics Board (CAB) in 1938, which began the era of formal airline regulation. Through the awarding of most airmail contracts to a select few companies, a process of massive consolidation ensued in the industry with the primary beneficiaries being American, Eastern, TWA, and United. The awarding of these contracts also laid the foundation for the dominant business model, which was based on government subsidies (until after World War II) for certain city-pair airmail routes that generally traversed the United States in an east-west pattern. The four main carriers accounted for more than 90 percent of all airmail service in the late 1930s.[27]

Although the fluidity of the early industry allowed the entry of a wide variety of strong personalities with deep experience to mold aviation, the next three decades told a very different story. This relative stasis in the industry contributed to the unusually long tenure of CEOs of the major U.S. airlines, averaging more than 30 years (see table I.2). Banking on the continued stability of the industry, most of the CEOs of the major airlines did not groom adequate successors. During this period, there was a strong focus on operational efficiency and standardization. In addition, the major airlines attempted to protect their positions through expanding the scale and scope of their activities—all within the defined regulatory business model of success. In essence, the government set the stage for the dominant business model during this phase of the airline industry's evolution, and the airline CEOs worked to solidify the model through investment in new technology and infrastructure. The scale and scope of these investments was a daunting inhibitor to new competitors who sought to enter the industry.

Table I.2 Tenures of initial chief executives of major U.S. airlines

Airline	*Executive*	*Tenure*
Pan American Airways	Juan Trippe	1927–1968 (served on board until 1975)
Delta Air Lines	C. E. Woolman	1928–1966 (died as CEO)
American Airlines	C. R. Smith	1934–1968, 1973–1974
United Air Lines	Pat Patterson	1934–1963 (chairman until 1966)
Eastern Airlines	Eddie Rickenbacker	1934–1953 (chairman until 1963)
Trans World Airlines	Jack Frye	1934–1947
Continental Airlines	Robert F. Six	1938–1980

Although established airlines grew during this period and their routes became more interconnected, the ability to adapt to major future changes in contextual factors diminished. As the Big Four worked to protect their dominance in the domestic airline market in the United States, Pan American under the leadership of Juan Trippe sought to carve out a sole, quasi-monopolistic business model for international travel originating from the United States. Trippe tried to protect his role as an "ambassador of the U.S." by expanding the destinations that his airline served and by investing heavily in the company's *international* brand identity, training, and infrastructure. Ultimately, Pan American's integrated investments in international travel inhibited its ability to adjust quickly to changing contextual factors, but its approach served the company well for four decades. We pick up the story of Juan Trippe and Pan American in chapter 4.

The tight regulatory environment that characterized the second phase of the airline industry's evolution gave companies few opportunities to differentiate themselves through pricing. Therefore, airlines sought competitive advantages principally through customer service and through new equipment (primarily larger, faster aircraft).[28] Even in this restrictive contextual landscape, leadership made a big difference. For most of this period, American Airlines' C. R. Smith developed and deployed innovations that the rest of the industry quickly imitated. From his promotion of higher safety standards through the development of the DC-3 in the 1930s to his venture with IBM in the late 1950s to create a much faster reservations system called SABRE (Semi-Automated Business Reservations Environment), Smith maintained American's reputation as an industry innovator. Smith's story will be explored in chapter 5. At the same time, other leaders emerged with different propensities focused more on improving internal systems relative to Smith, who was more outwardly focused on the customer. United Air Lines' Pat Patterson, for example, was deeply influenced by his humble early career and strived to build an airline characterized by progressive people-oriented policies. He also worked on enhancing communication between different segments of the company (e.g., corporate headquarters and operations) as well as initiating a state-of-the-art maintenance facility following World War II. Patterson's approach to maintaining the dominant business model in the airline industry will be covered in chapter 6.

During this second phase, some of the most important changes in the contextual factors influencing the growth of the aviation industry were in the realm of global events and consumer acceptance. The improving safety of commercial aircraft served to allay the public's long-held fears about flying. As a result, international and domestic tourism grew very quickly: between 1938 and 1954, the total revenue-passenger miles U.S. airlines flew increased from 480 million to 16.7 billion and the total number of airline passengers increased from 1.1 million to 32.2 million.[29] The supremacy of the Big Four was reflected in their continued dominance in terms of revenue-passenger miles, which even at the end of this extraordinary growth period remained as high as 71 percent (starting from an even more dominant 82 percent share in the beginning).[30] While the Big Four dominated the airline industry with a similar business model, smaller airlines sought creative ways to establish a competitive foothold. Regional airlines such as Braniff Airlines in Texas and Pacific Southwest Airlines in California took advantage of less stringent intrastate regulations to build viable companies during this phase. In addition, Delta Air Lines was successful during this period by primarily serving a less trafficked sector of the country.

The inflection period between the second and third stages of the airline industry's evolution lasted ten years, pulling the industry from an era of relative certainty or predictability to an era of increasing uncertainty.[31] By 1968, the business model that had moved the airline industry from great volatility to relatively more consistent profitability had been entrenched for many years. Regulation and the national route structure favoring the Big Four had continued without major alterations for 30 years. The CAB generally reviewed route requests on a case-by-case basis and preferred creating point-to-point routes.[32] This was hardly the most efficient way to organize the industry's national route system, but the major airlines were left with the task of investing for years in servicing and maintaining this system. Airline CEOs neither advocated major innovations within the regulatory regime nor did they push the government to change the regulatory regime itself.[33]

These largely self-inflicted wounds were aggravated by shocks in the economy and international relations. The spikes in international terrorism and especially in oil prices along with the recessions of the 1970s conspired to rudely change the stable contextual landscape for the airline industry. The 1970s were also a moment when many politicians and economists in the United States became dissatisfied with regulations of many industries, not just airlines. The long-time government regulations that had controlled railroads, telecommunications, trucking, and the financial industry, for instance, all came under attack. In the case of the airline industry, many critics of regulation pointed to the success of intrastate airlines—such as Pacific Southwest in California and Southwest in Texas—that prospered well out of the protective federal umbrella of the CAB.

In short, those key elements that had coalesced to stave off changes in the dominant business model of the airline industry—a lack of turnover in leadership and a stable longstanding regulatory regime—were fraying due to the erstwhile cumulative changes in legal, economic, and international conditions, as well as the mortality of the airline leaders themselves. Leadership and regulations could not stave off these changes indefinitely, but their success over the previous 30 years made it all the more difficult for the dominant airlines to adapt.

Furthermore, promising new business models for the airlines, such as Southwest, had made money by directing their operations and their investments toward a very different market than traditional airlines. Southwest's model of success was in place almost a decade before deregulation took hold in the airline industry, and its business model under the leadership of Herb Kelleher will be explored in chapter 7.

The uncertainty of this second inflection stage increased as a result of three primary factors: (1) the unusually high costs incurred by the dominant players during the race to buy the latest generation of aircraft in the late 1960s; (2) the high turnover in executive leadership; and (3) the highly integrated investments in the prevailing industry infrastructure. In his study of regulation in American industries, Richard H. K. Vietor noted that "regulation induced excess capacity, caused higher-than-necessary costs, retarded innovation, and severely distorted patterns of supply and demand."[34] In summary, the industry was characterized by increased complexity, which made change very difficult.

Phase III—Postregulation: A New Business Model for Success

Much like in its first phase, the third phase of the evolution of the airline industry was characterized by a very wide range of experimental business models that emerged after deregulation in 1978. Despite their many failures, the presence of new entrants fighting for market share plunged the industry into a series of price wars. Carriers were often confused about which business model to pursue. For example, Continental in the 1970s and Delta in the early 2000s attempted to maintain their old business model while simultaneously starting airlines-within-an-airline (Continental Lite and Delta Song, respectively) to compete with Southwest and other low-cost providers. Sadly, this strategy did not work and eventually had to be disbanded, leading to even greater losses for these once dominant airlines.

While the deregulation of the airline industry in 1978 posed great problems for many of the large established airlines, the fortunes of new airlines entering the national stage at this time were hardly uniformly successful. Many new entrants became overconfident with their ability to challenge established competitors so quickly. Donald Burr's low-cost and lower frills People Express made a huge splash with the American public as well as the media (becoming at the time the fastest company to rise into the ranks of the *Fortune 500*) but fell to a quick and ignominious end after Burr attempted to go head-to-head with "the majors."[35] Unfortunately for the aspirations of postderegulation entrants, People Express was just one of the most visible of a long series of failures. According to the Government Accountability Office, there were 160 airline bankruptcies between 1978 and 2005.[36] Figure I.2 shows the consolidating and widening impact of regulation and deregulation in the airline industry.[37]

By the 1990s, the stability that had characterized the long second phase of the airline industry had given way to chaos. Fierce competition in ticket pricing made cost-cutting a huge priority. Some airlines did this by concentrating

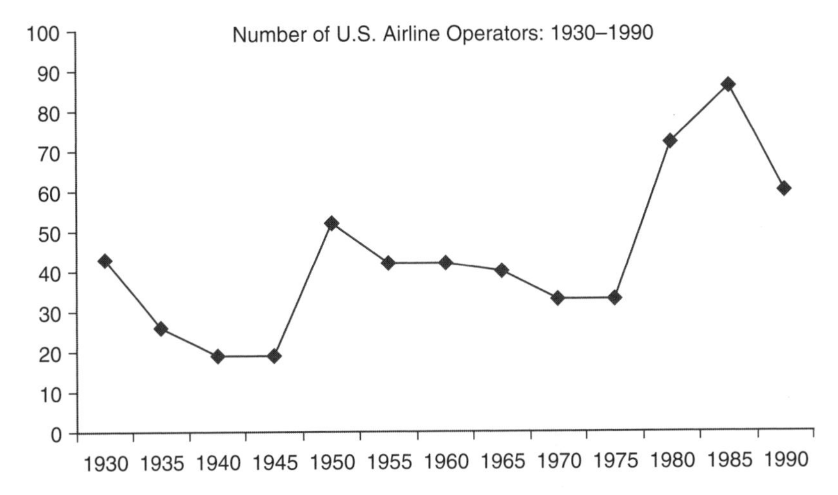

Figure I.2 U.S. airline industry consolidation and expansion from 1930–1990

their operations in huge hubs, such as American Airlines in Dallas and Delta in Atlanta; others did this by drastically cutting their labor costs, as did Continental Airlines under Frank Lorenzo. We will explore the demise and rebirth of Continental Airlines in chapter 8 under the leadership of Gordon Bethune. In this chapter, we will also finish our story of Pan Am after Trippe resigned from the CEO post as an example of a company whose great historical success contributed to its difficulty to change when the times demanded it.

By the mid-1990s, Kelleher and his Southwest Airlines had emerged as the darling of the industry—the only major carrier to make a profit year after year. Other airlines, such as American, worked hard to maintain their market share, but their profits generally remained thin. The decision to invest in a hub and spoke system in the early years of deregulation committed many of the majors to maintaining a long, nation-wide route system stitched together by connecting flights. Although it delivered savings in operating expenses in the short-term, the hub and spoke model was much more complicated than point-to-point travel and therefore subject to more potential problems and delays.

Another Jolt to the Industry, Post-9/11

In the late 1990s, rising fuel prices constantly undercut the older airlines' efforts to find profits through reducing costs. In 1999, jet fuel prices went up 29 percent. This dismal scenario for U.S. airlines played out in the first quarter of 2000 when profits for *all* the major airlines amounted to a measly $79 million.[38] In recent years, before *and* after September 11, 2001, low-cost airlines consistently outperformed the older airlines in terms of lower costs per seat mile by 20 to 30 percent.[39] Older established airlines plummeted billions into the red while low-cost airlines remained in the black, albeit by a hair.[40]

The major airlines' plea for government aid in the wake of the 9/11 terrorist attacks on the United States may have been a legitimate response to a horrifying situation that put a deep chill in airline travel, but it may have also masked the

Table I.3 U.S. airline industry profitability before and after regulation and 9/11

	1938–1978	*1979–2001*	*2002–2006*
Cumulative Operating Profit/Loss	$10.8 billion	$44.8 billion	($4.2 billion)
Cumulative Net Profit/Loss	$5.4 billion[a]	$4.4 billion	($23.6 billion)

[a]The cumulative net profit figure reflects data for 1947–1978.

fact that the older airlines had already begun in the year 2000 to use loans to bolster their cash balances—a trend that has continued to the present day.[41] Between 1938 and 2002, the U.S. airline industry produced a cumulative loss of $1.1 billion. By 2006, the cumulative loss exceeded $13 billion (see table I.3).[42]

New changes in technology—a contextual factor that has not changed much in recent decades—such as ultralight jets, could further devastate the entire industry. Businessmen, who have generally been willing to pay high fares for last-minute travel arrangements, may use these little jets (much smaller and cheaper than traditional corporate airplanes) as low-cost alternatives to today's airline choices.[43] If enough businessmen decide to make the switch, what would become of existing airlines that depend on them so much? And what happens to the industry if fuel costs continue to skyrocket? What role should the government play going forward? What type of leadership will be necessary at this stage in the evolution of the airline industry?

The reciprocal relationship between context and leadership that is evidenced in the century-old history of the airline industry is representative of general trends in any industry's evolution. As industries progress from start-up to maturity to rebirth, the nature and role of leadership must change and evolve. In this book, we explore the evolution of an industry through the actions and decisions of leaders who shaped and were shaped by the context of their times. Specific leadership archetypes (entrepreneurs, managers, and leaders) generally govern or dominate particular lifecycle stages of industries (start-up, maturity, and decline). Success is often contingent upon the right leadership approach for the situation at hand, but the relationship between leadership style and situation is not one dimensional. The situation can influence the leadership that is appropriate and necessary, yet business executives can also influence and alter situations to fit specific goals and objectives. This book explores this co-evolutionary process of industry development through the stories of entrepreneurs, managers, and leaders in the airline industry. The airline industry has its own set of idiosyncrasies but so do all others. Because the interrelationship between its evolution and the role of leadership is representative of other industries, the lessons which can be drawn from the detailed portraits of airline leaders in their times will enable readers to better assess the interplay between context and leadership within their own industries.

PART I

The Entrepreneurs

The initial, start-up phase of any industry is typically characterized by chaos, uncertainty, risk, and experimentation as entrepreneurs seek to create a sense of legitimacy for their operations. The vast experimentation during this period takes the form of a diverse array of business models, each one scrambling to become dominant. In many cases, the individuals and their approaches are years ahead of their time, and as such, it is no surprise that start-up phases within new industries are fraught with many failed business attempts. More than at any other time in an industry's evolution, however, the role of the individual actor/entrepreneur is vitally important. Entrepreneurs make investment decisions, create business plans, allocate scarce resources, galvanize followers, and articulate a vision for a future state. While external forces such as demographic shifts or government regulation or geopolitical forces can create the conditions for the development of new businesses and industries, it is up to the individual actor to bring the disparate pieces together. As businesses and then industries emerge, entrepreneurs have a disproportionate influence on the early evolutionary forces. In essence, they create the platform and context for success.

Consider the early experiments to create viable businesses using the Internet as the backbone. In the early days of the Internet, hundreds of businesses emerged predicated on various business models—some using the Internet as a retail portal, others using it as a destination or community site, and still others using it to round out a portfolio of distribution options. While many Internet-based companies were created without a realistic business proposition and evaporated into cyberspace, a handful of businesses like Amazon, e-Bay, and Google were built that have defined success in the Internet industry. Like other successful pioneers, Jeff Bezos of Amazon, Meg Whitman and Pierre Omidyar of e-Bay, and Larry Page and Sergey Brin of Google created business models that have been copied, molded, and adapted by others to varying degrees of success. The initial shake-out of the industry has run its course, setting the stage for the next phase of growth. Almost one hundred years earlier, the "silicon valley of entrepreneurship" was not in California, but in Battle Creek, Michigan home of Postum Cereal Company and Kellogg's. Introduced at the turn of the twentieth century, Post's Grape Nuts and Kellogg's Corn Flakes revolutionized the instant breakfast cereal industry and sparked a wave of innovation and imitation. By 1902, Battle Creek was home to more than 35 cereal companies attempting to tap into the country's growing desire for greater convenience and service by introducing their own versions of breakfast cereals made from grapes, wheat, and even celery.[1] Although many of these companies did not survive, the ones that did developed a business model focused on national advertising, expanded distribution, quality and safety, and competitive pricing.

The airline industry is also emblematic of this process. Entrepreneurship is not a one-dimensional pursuit; it can take many forms and approaches. Early entrepreneurs sought to legitimate the industry through a variety of business models—from government service for the U.S. Postal Department to agricultural business expansion to passenger service to international diplomacy. In the first part of this book, we explore the approaches of three entrepreneurs in the early years of the airline industry who sought to develop a viable business model that would be both sustainable and profitable. The three entrepreneurs whose stories we tell in this section represent three prototypical types of entrepreneurs—*foundational entrepreneurs* (Harry Guggenheim) who work to create the institutional structures and foundation for success; *frontier entrepreneurs* (Juan Trippe) who operate at the edge and set the pace for what can be and what is possible in a new industry; and *fast-follower entrepreneurs* (C. E. Woolman) who quickly capitalize on a budding opportunity.

We begin with the story of Harry Guggenheim who is an exemplar of the foundational entrepreneur. He worked to create the conditions for the success of other entrepreneurs in the airline industry by focusing on institutional policies and securing resources for the new industry. Guggenheim strived to popularize the notion of safe, passenger air travel decades before it became a viable business proposition. Through his efforts to fund research on technology and aircraft safety, Guggenheim was determined to eliminate one of the largest obstacles to the viability of the uncertain airline industry—consumer fear.

Unlike many other entrepreneurs, Guggenheim wanted to share ideas, technology, and approaches across competitors—both aircraft producers and the early airline companies. He believed that the sharing of best practices would speed the process of building legitimacy for the airline industry and would ultimately accelerate the pace of consumer acceptance. His efforts to build a community of research helped to begin this process, but in many ways, he was well ahead of his time. Passenger travel would not be economically viable until after World War II when investments in technology (spurred by the productivity increases for the war effort) resulted in the development of safe and cost effective means of air travel. Guggenheim is not unlike many early pioneers in an industry who helped to create the stage for success without necessarily directly benefitting from their work.

In contrast, Juan Trippe, the founder of Pan American World Airways, worked throughout his career to create, protect, and defend a government-supported monopoly on international air travel for decades. During his early years, Trippe consistently operated at the edges of the frontier, which initially brought mixed results. His first of several efforts to build a domestic airline system by securing point-to-point contracts for airmail resulted in failures. He was initially so enamored by new technology possibilities and rapid expansion opportunities that he overextended his airline's financial resources. Though he failed to build a domestic airline, Trippe learned two very important lessons—the ability to strike first and the significance of lobbying efforts. Locked out of the domestic market by others, Trippe lobbied hard for the first international airmail contract. After winning the airmail contract from Key West, Florida to Havana, Cuba in 1927, Trippe pursued a path toward monopolization by swiftly and deftly securing landing rights to several Latin and South American destinations. He moved quicker than anyone else and simultaneously maintained a steady stream of lobbying in Washington, D.C. to secure his dominance, playing into the country's desire to maintain a strong and decisive influence in the Western Hemisphere.

Trippe was so successful in creating an international airline that Pan American was considered a form of American diplomacy throughout the world for almost four decades, but he was, perhaps, too successful. Trippe was so focused on the protection of his quasi-monopoly that he failed to adequately recognize and prepare for the changes in the contextual landscape. As his dominance of international travel was challenged by others, Trippe and Pan Am were unable to respond appropriately and the company eventually ceased to exist. We will follow the success and ultimate demise of Pan Am throughout the book.

While Guggenheim sought to popularize the notion of air travel for the public and Trippe created a dominant presence in the international arena, C. E. Woolman looked for an opportunity to create another business model within the airline industry. Like almost all industries, there were countless opportunities and vast untapped potential in the early phase of the airline industry. Combining his background in agricultural management with his fascination with flying, Woolman began Delta Air Lines as a crop dusting business in the farming rich area of Southeastern United States and quickly expanded to regional passenger service from Dallas, Texas, to Atlanta, Georgia. Starting small and focusing on a relatively underdeveloped industrial sector of the country, Woolman was not an initial player in what was considered the only viable means for steady income from flying—domestic airmail. When government forces created an opening for new airmail routes, Woolman, as a fast follower, jumped on the opportunity. By the time he secured airmail contracts, Woolman had already built a strong regional passenger business. In turn, Woolman's regional business model would be followed and mimicked by many others (Braniff and Southwest) as they sought to gain a foothold in the airline industry.

The approaches of Guggenheim, Trippe, and Woolman were just three of many that were tried by erstwhile aviation enthusiasts. Given the seemingly limitless potential, there was very little overlap between business models. With no standard technology, no organized route and safety system, no initial government influence, and no consumer demand, the opportunities within the early airline industry were wide open, if not entirely designed for quick profitability. Entrepreneurs experimented with a variety of business models to find one that held both promise and potential. As their businesses gained traction and support, a flood of "me-too" companies were created to ride the bandwagon, setting the stage for the next evolutionary phase in the industry's development. As we will see in Part II, the move from start-up to stability in the industry, which was championed by early entrepreneurs, was significantly reinforced through the actions of government.

CHAPTER 1

The Guggenheims: Promoting Aviation in America

In October 1929 the American press, both large publications and small, began to bid farewell to the Daniel Guggenheim Fund for the Promotion of Aeronautics, which had just made an announcement that it would soon cease the operations it had inaugurated three years earlier. The *New York Times* began its tribute by proclaiming that aviation itself had become "the flourishing protégé of Harry F. Guggenheim and his associates."[1] In a November editorial from the lay Catholic magazine *Commonweal* titled "Aviation Is Weaned," the Fund was lauded for having helped "American aviation off to the flying start it should properly have."[2] This was no mean feat. In essence, after having distributed approximately $4,000,000 during the previous five years, Daniel and Harry Guggenheim had provided the timely financing and the vision necessary to revive the moribund American aeronautics industry. Certainly, the Guggenheim organization was not the only entity to set its sights on improving American aviation—especially commercial and passenger aviation—in the second half of the 1920s. Nonetheless, *Commonweal* explained, "in any discussion of airways progress in the United States, the work of the Guggenheims is the obvious thing to start with"; its achievements were so impressive, the editorial continued, that it "is more likely that the importance of the Fund will be overestimated than underestimated."[3]

The editors of *Commonweal* were half right. The contributions to American aviation made by the Guggenheim fund—financed by Daniel Guggenheim and guided by his son Harry—were breathtaking in scope. Their money provided crucial support and direction to the theoretical, practical, and public-relations problems facing commercial aviation in the United States. With remarkable vision and leadership, Harry Frank Guggenheim practically took it upon himself to integrate aviation (and the efficiencies and new markets it promised) firmly into the American economic system by attacking a fundamental problem: the real *and* perceived dangers of mechanized flight. More than anyone of his time, he recognized the twofold challenges posed by airplane safety. In "Creating Air-Wisdom in the Public," he wrote that those "who are deeply interested in the progress of aviation find themselves concerned with two things[:] the constant improvement and perfection of the airplane and its facilities for navigation; and the public knowledge of these developments."[4]

But, as of 2009, Harry Guggenheim is hardly a household name associated with aviation in America. Even a recently published Smithsonian guide to the history of aviation has left Guggenheim out of its index. Interestingly, many important names in aviation history that were closely associated with Guggenheim do make an appearance in the guide, including Commander Richard Byrd and especially Charles Lindbergh. James Doolittle, the daring navy pilot, is remembered for breaking speed records in the 1920s as well as a harrowing bombing raid over Tokyo in 1942; nonetheless, his role in successfully flying and landing the first airplane in zero visibility—a project sponsored by the Daniel Guggenheim Fund for the Promotion of Aeronautics—is left unmentioned. And even President Herbert Hoover, who very publicly endorsed Harry Guggenheim's Fund as Secretary of Commerce and who announced Guggenheim's appointment as ambassador to Cuba in 1929, is mentioned only briefly for his efforts to improve aviation that went back to the early 1920s—efforts that anticipated Guggenheim's work in the second half of the decade.[5]

Why have the Guggenheims remained obscure in America's aviation history? Most aviation literature has been attracted to the spectacular developments of airplanes rather than the relatively more mundane task of how the complex scientific and economic problems of the aviation industry have been solved. According to historian Dominick A. Pisano, histories of aviation are "infused with enthusiasms of all kinds, but especially for the artifact [i.e., the airplane and its wondrous capabilities] over other important considerations."[6] In addition, Guggenheim was not a leader who sought publicity for himself. Although he was not afraid of the limelight, Guggenheim stepped into its glare mainly as a spokesman for American aviation and attempted to share that limelight with colleagues whenever he could. As an advocate and spokesperson for the industry, Guggenheim argued that an airplane's unique ability to overcome geographic barriers could promise much more: it offered "the opportunity for intimate contact and better understanding that points the way to a world empire not based on the ephemeral military supremacy of a Rome, but through the real civilization of mankind."[7] Guggenheim's efforts are characteristic of entrepreneurs who establish new foundations. Before many others, Guggenheim saw the airplane as a vehicle for facilitating broad global understanding, diplomacy, and commerce. He had a rare ability to see a future vision of the airline industry in America.

Early Aviation Industry in America

Many American commentators in the 1920s remarked with great disappointment that although the United States could boast of having been the birthplace of mechanized, heavier-than-air flying machines, aviation in America lagged far behind Europe. This disparity had occurred despite the fact that, as reported by the Department of Commerce, the United States—with its huge population, landmass, and economy undivided by the many national borders that honeycombed Europe—was the ideal venue for the growth of commercial aviation.[8] The obstacles restricting the progress of American aviation were complex and had long histories. Although Wilbur and Orville Wright had catapulted America to its position as the premier aviation country in the world, which a stunned international audience witnessed during the Wrights' 1908 demonstration flight in France, the Wrights would, ironically, condemn the United States to aviation mediocrity in the 15 years that ensued. The Wrights' slow and meticulous methods—which were insulated from the world while they were making five-years' worth of improvements in their Ohio workshop after their success in Kitty Hawk in 1903—were poorly adapted to the more frenetic pace of technological change spurred by competitors and imitators from all over the world. Instead of innovating and improving on their original invention, the Wrights spent many years attempting to protect their design from competitors, abroad and at home, by claiming that other airplane manufacturers had infringed on their patents.[9]

Some of their fiercest American competitors, most notably the famous aviator Glenn Curtiss, were much more consumer-oriented than the Wrights and focused on making technical and design modifications to airplanes that appealed to the small but growing number of Americans interested in flying their own planes.[10] The many patent lawsuits initiated by the Wrights against Curtiss thwarted this promising competitor who often responded by making minor modifications to his airplane design. This, of course, provoked more complaints from the Wrights. The small group of Americans with some interest in aviation in the early 1910s divided into opposing Wright and Curtiss camps and thus missed their opportunity at becoming a fraternity whose main interest would be the growth of the entire industry. Henry Ford actually instructed his own lawyers to help Curtiss in what *Flight* magazine called the "aviation war." Ford argued that patents "don't . . . stimulate invention . . . but they do exploit the consumer and place a heavy burden on productive industry."[11] This legacy of distrust characterized the U.S. aviation industry for years to come. The Wrights remained competitors in the courts (instead of in their workshop) until Orville became exhausted with the struggle to stem the tide of innovation; Orville Wright sold his company in 1915 (Wilbur had died in 1912).[12]

Although their litigious nature stalled the development of the airline industry in the United States, the Wrights' accomplishments in aviation in 1908 helped reenergize the burgeoning aviation industry in Europe. The French were especially motivated. Their pride in aviation antedated that of the United States by 125 years when, on November 21, 1783, a balloon designed by Etienne and Joseph Montgolfiere first lifted two men into the sky without a tether.[13] France's ability to catch up to the Americans was ably demonstrated as early as 1909 when Louis Bleriot successfully piloted one of the first monoplanes across the English Channel.[14] These giant steps forward in aviation coincided with a growing European arms race that led up to World War I. As one historian has

Wilbur Wright (left photo) checking biplane. Glenn Curtiss (right) at controls of biplane. (Source: *The Harvard-Boston Aviation Meet at Atlantic, September 3 to September 13, 1910,* photograph album. Baker Library Historical Collections, Harvard Business School).

pointed out, an arms race among nations in such close proximity made improvements in aviation not only a military necessity, but also a point of national pride. In general, the European public shared the enthusiasm of their generals who understood that aviation had great military potential.[15]

Harry Guggenheim saw these contrasting transatlantic trends in aviation first hand. As a young man, Harry decided to join the war effort as an aviator. He did this as a pilot flying for the U.S. Navy's Northern Bombing Group in the Camproni, a heavy bomber made in Italy. Indeed, Harry was far more likely to fly planes made by European manufacturers than by American companies. Famous American aviators flew the French Nieuport and SPAD or the British SE.5a, built by the Royal Aircraft Factory.[16]

Even though the U.S. government had pledged to fill the European sky with American aircraft after declaring war on Germany in April 1917, U.S. aircraft manufacturing at this time was an almost unmitigated flop. By the end of the war, the United States had supplied only 2.5 percent of all the planes manufactured for the war effort.[17] At least, the declaration of war had motivated rival aircraft companies to agree to cooperate to some degree. Instead of fighting over patent rights, patents were shared and manufacturers paid a fee to use particular patented designs. But there were other problems facing the growth of airplane production.

Many naïve manufacturers—jumping blindly into the aviation field as Congress was allotting hundreds of millions of dollars for airplane manufacturing—thought that the methods used for making cars could be replicated for airplanes. Unfortunately, the largely unskilled labor pool that enabled the huge growth in automobile manufacturing could not be tapped for the war effort. Because airplanes of the day were so fragile and unstandardized, laborers had to work to the level of highly skilled artisans. The government further complicated this situation by often changing the specifications of airplanes. Meanwhile, the steady gains in airplane design made by European manufacturers meant that even when an American design actually was completed on the factory floor, those airplanes were often obsolete when they were introduced to the battlefield. Immediately following the war, the glut in military surplus aircraft in the United States undercut the efforts of manufacturers to transition from wartime to peacetime requirements. The only clear advance made in U.S. aircraft production during the war was the development of the Liberty engine, a reliable machine capable of producing 400 horsepower.[18]

Officials from the U.S. government became alarmed by the aviation strides made in Europe that had been facilitated by European governments' willingness to subsidize their fledging aeronautics industry—from manufacturing firms to airline companies. Americans were not willing to follow the European example, but they did see a role for government to help regulate this untamed industry. This approach was shaped largely by President Warren G. Harding's young Secretary of Commerce, Herbert Hoover, who joined the administration in 1921. Hoover had gained a sterling reputation for his brilliant direction of the effort to distribute food all over war-ravaged Europe in the immediate aftermath of World War I. The Guggenheim family admired Hoover's abilities and asked him to become a senior partner of their hugely successful mining business with an annual salary of $500,000. After thinking it over for a week, Hoover turned down the offer in favor of joining the government.[19]

U.S. Government's Role in Early Aviation Industry

Upon Hoover's entrance into Harding's cabinet, commercial aviation in the United States was largely under the regulatory aegis of the Department of Commerce (although airmail was directed by the post office). Widely celebrated by the American press in the 1920s, Hoover was the embodiment of the can-do technocrat who had deep faith that American business would develop other can-do technocrats who shared his vision and his ability to run business at maximum efficiency for the good of the whole population. Hoover hoped to shape American commerce through his vision of the "associative state" in which the government would offer anything but direct monetary assistance to any given segment of business or industry. In the case of the civil aviation industry, Hoover hoped to spur growth through sharing technical and economic research to improve the management of airlines. In addition, Hoover's Department of Commerce tried to shift public perception about airplanes, which was seen as the domain of daredevil "barnstormers:" a group of predominantly ex-military pilots who eked out a living going town-to-town performing stunts and offering rides for the stout-hearted. Instead, Hoover hoped to rescue the reputation of mechanized flight in America by convincing the public that aviation could play a positive role in the economy by raising profits through increased efficiencies in transportation.[20]

Unfortunately for Hoover, the aviation industry lacked any vibrant voluntary trade association on which the "associative state" depended. Ideally, these trade associations would perform essential functions such as setting voluntary industry safety standards for workers and consumers. But, in the early 1920s, American aviation was still in disarray with no central strategy or direction emanating from either government or industry. Stunts and accidents still dominated the headlines, relegating aviation to the backwaters of the American economic landscape. During this time, even the famous American humorist Roy Rogers recognized that aviation was better at getting laughs than gaining profits, an observation that turned into a grim joke: "Five people killed in a plane yesterday and it is headlined to-day in every paper. Saturday in Los Angeles at one grade crossing seven were killed and six wounded and the papers didn't even publish the names. It looks like the only way you can get any publicity on your death is to be killed in a plane. It's no novelty to be killed in an auto any more."[21]

It was at this moment that Harry Guggenheim entered the frayed economic and cultural nexus of American commercial aviation. For Harry Guggenheim, as well as his father Daniel, it was also a time of transition for their family. Just a few years earlier, Daniel Guggenheim prepared for his retirement by successfully selling the family's Chilean mining business for the sum of $70,000,000 (approximately $707 million in 2000).[22] Harry had worked at the family business for a few years: from 1908 to 1910 at American Smelting and Refining Works in Aguascalientes, Mexico and later as a partner from 1916 to 1923 of the Guggenheim Brothers firm. In between, he earned B.A. degrees in political science and economics at Cambridge University in England studying under none other than John Maynard Keynes. (He later earned an M.A. at Cambridge in 1918.) In 1924, Daniel Guggenheim's wealth and long-held interest in philanthropy was channeled into the Daniel and Florence Guggenheim Foundation whose broad mandate was to promote "the well-being of mankind throughout the world."[23]

Guggenheim School of Aeronautics

A few months after the foundation was created, Harry Guggenheim was asked by the chancellor of New York University (NYU) to join a committee with five other men whose mandate was to start a campaign to raise $500,000 to establish a school of aeronautical engineering. Harry Guggenheim responded to the chancellor's invitation as a welcome opportunity to follow his father's footsteps in philanthropy and to cultivate further their shared interest in the future of aviation. As early as 1918, Daniel Guggenheim expressed this faith in a letter to his son in this comment about World War I: "I wonder whether you think as much of the aeroplane as bringing the war to a final end as I do?"[24] At first, the six-man committee favored soliciting the public at large for funds. But Harry Guggenheim was keenly aware of public sentiment, which was hardly enthusiastic about the prospects of aviation in mid-1925. Instead, Harry proposed that the chancellor write a letter describing the merits of NYU's proposed aeronautics school for individual donors; furthermore, he volunteered to deliver the letter himself to his father and uncles first. After Daniel Guggenheim read the letter, he responded: "Don't show this letter to your uncles, Harry. I will do it myself. I have given all my life to work underground; now let me see what I can do to help above ground."[25] On June 15, a public announcement was made concerning a $500,000 grant to NYU given by Daniel Guggenheim. Approximately half of the grant was allocated to buying scientific materials—including a wind

Daniel Guggenheim breaking ground for the N.Y.U. School of Aeronautics. (Source: Bettmann/CORBIS).

tunnel—and the other half was used to establish three chairs in aeronautics, along with some lab assistants.

In the immediate aftermath of the Great War, most military officers favored naval over aerial defenses because airplanes had not proven to be decisive in the war's outcome.[26] General William "Billy" Mitchell, along with allies like Pan American's founder General "Hap" Arnold, had vociferously defended the potential of aviation and advocated uniting the military's air forces under one command (instead of being divided between the army and the navy). Eventually, Mitchell's impatience with the future of aviation being hampered by people who had little or no personal experience with airplanes moved him, by at least 1925, to become a vociferous advocate of "a unified department of aeronautics" that would regulate both military *and* civilian aircraft.[27] In that year, Mitchell took every opportunity to dominate the headlines by attacking what he saw as the malfeasance of the air divisions of the army and navy.[28]

In response to Mitchell's attacks, President Calvin Coolidge organized a special commission on aviation directed by Dwight Morrow, an old college friend of Coolidge's and a partner of J. P. Morgan since 1914. In November, Morrow's committee submitted a report to the president recommending that the Department of Commerce set up a Bureau of Aeronautics whose role would be mainly to bring some order to the chaos of American aviation. It would "regulate civil air navigation . . . , license pilots and inspect aircraft, maintain air routes and air navigation facilities, regulate international civil aviation as it affected the United States, and encourage and promote the growth of civil air transport service." Most of these recommendations were integrated into the Air Commerce Act, which was passed into law on May 20, 1926. In addition, throwing a sop to the supporters of Mitchell's efforts to reform military aviation, the Morrow commission recommended that the role of airplanes be strengthened in the military through a new post of "assistant secretary for aeronautics" that would be added to the commerce, navy, and war departments.[29]

For those who believed that the genius of the American economic prosperity and even its cultural vitality lay in the ability of private individuals and industry to exercise a great deal of autonomy and freedom (especially compared to the European economic model), the United States had certainly dodged a bullet. Although the Air Commerce Act of 1926 succeeded in bringing some order to the world of American aviation, the Coolidge administration could not have acted much more aggressively to rescue civil aviation without discrediting the very "associative state" model that it had long promoted and more recently defended against critics such as Mitchell.

Establishment of the Guggenheim Fund for the Promotion of Aeronautics

As the Mitchell controversy crested in the fall of 1925, Harry Guggenheim continued to explore ways to advance the aeronautics industry in the United States. A friend who worked in public relations, Ivy Lee, suggested to Harry that he start a "fund for the promotion of aeronautics." Finding this suggestion intriguing, Harry sought the input of many friends and colleagues who concurred that this initiative had promise. Soon afterward, Harry asked the counsel of none other than Orville Wright who also supported the idea. With Orville Wright's support, Harry found no difficulty in getting his father to promise $2,500,000 for a fund with a very broad mandate in the field of aviation: "to sponsor education, research and development, and aviation promotion via publications, brochures, and publicity demonstrations of safe flying."[30]

Daniel Guggenheim, whose age and frail heart forced him to take a back seat to his son in matters of the day-to-day planning and operations of the Fund, still provided far-sighted counsel. Sensitive to the controversies surrounding aviation brought about by Mitchell's public diatribes, Daniel Guggenheim argued that further political storms could be provoked by Harry's initiatives: "To be of any value," the elder Guggenheim counseled his son, "our Fund must be tendered to the government. With all the hue and cry that's going on, we'd better make certain that the government will accept it."[31] As might be expected, the Guggenheims enjoyed close connections with the power elite, including Dwight Morrow, who had just finished his appointment by President Coolidge as chairman of the commission that made recommendations to the president concerning aviation in the wake of the Mitchell controversy.[32]

Morrow arranged for Harry Guggenheim to have an audience with the president. After a quickly arranged lunch meeting with Guggenheim, Coolidge nodded his approval of the Fund's proposed activities.[33] With this taciturn blessing from the White House, Guggenheim soon afterward gathered an impeccably credentialed board of directors that included stars from science, finance, and aviation to help him to direct the Fund.[34] On January 16, 1926, the Guggenheim Fund officially began with a letter to Secretary of Commerce Hoover announcing its intention to "further the application of aircraft in business, industry, and other economic and social activities of the nation."[35]

The problems and possibilities confronting aviation at the time were many, and Harry Guggenheim responded with a commensurately broad agenda. There were tentative plans to fund research in the development of "helicopters, radio direction finders and aerodynamics." Harry also proposed the idea to sponsor a competition that would offer a large grand prize for a commercial plane that could make marked improvements in "safety and stability, and . . . improved engine design." Another important task would be a public education program that attempted to balance what Harry Guggenheim took to be the lopsided negative depiction of aviation in the popular press.[36] The first undertakings the Guggenheims set out for themselves and their fund, however, were to continue their efforts to increase aeronautics education in the nation's universities and to gather the most up-to-date information on aviation in 1926. To accomplish the latter task, Harry Guggenheim traveled to Europe.

From February to April of 1926, Harry Guggenheim, accompanied by Rear Admiral Hutchinson I. Cone who had commanded American naval air forces during World War I, toured Europe to obtain a detailed look at recent strides in aviation. After the two men had interviewed approximately 100 aviation leaders from most of Western Europe, Guggenheim reported to the press that Germany,

England, and France were particularly advanced in the fields of commercial aviation, aviation design, and airplane manufacturing, respectively, and that U.S. aviation could learn from these examples and catch up within five years.[37] Although "Government subsidies explain Europe's progress," Guggenheim posited, he remained "utterly opposed to subsidies here." The most the U.S. government should do was to establish a bureau that could provide "meteorological and communications services" for airplanes. Asked about the problem of encouraging American funding for an industry that "could not bring profit for a long period," Guggenheim expressed no worries. As the *New York Times* reported, "he held that the field in this country was so vast and promised such rich returns when developed that initial losses were justified."[38]

Focus on Safety

Guggenheim's exploration of Europe soon inspired him to unite the Fund's myriad efforts behind one overarching theme: safety. Of particular interest to him was recent research and development that had the potential in the near term to "reduce aircraft landing speeds and increase airplane stability." As Guggenheim Fund historian Richard P. Hallion has explained, technical problems that dogged airplanes in the 1920s were largely derived from engine failure and high stall speeds.[39] The first problem would be largely solved by encouraging the further improvement of the trimotor airplane that could, in times of crisis, stay aloft for a while even if two of its three engines broke down.[40] During landing, high stall speeds (the speed at which an engine would stop running) also caused crashes for airplanes that were forced to land in relatively small areas, such as a farmer's field. These high landing speeds also required airports to be equipped with long runways, thus increasing the initial expenses for municipalities that might be interested in encouraging aviation.

Although these problems were difficult to surmount, the European advances in aeronautical research convinced Guggenheim that marked improvements in airplane safety could be made in the next few years. By June 2, 1926, Guggenheim had convinced the board of his aeronautics Fund that the organization should "adopt as its primary policy 'the Promotion of Safety in Aviation' and that it concentrate its efforts towards every practicable means to accomplish this purpose." That same day, the board approved Guggenheim's suggestion that the Fund sponsor an international "safe plane" competition with $150,000 to $200,000 in total prize money.[41] Although the Fund did not officially announce the rules and dates of the competition until almost a year later, the *New York Times* reported on the plans for such a competition, noting: "Daniel Guggenheim Says Efforts Will Centre on Safety at First, to End Fears of Flying."[42] With characteristic candor and an ability to make complex subjects understandable to the American public, Guggenheim explained the problems confronting aviation:

> Today, when the speed of the conventional airplane is reduced below its so-called "stalling speed," it ceases to function like an airplane. If the airplane is flying at sufficient altitude and is aerodynamically well constructed, the pilot may recover

> from the "stall" and regain control. However, should the stall occur in leaving the ground, the cause of a great number of accidents, a crash is inevitable.[43]

With much less fanfare, Guggenheim ended the article by enumerating the Fund's related safety goals that would not be addressed in the competition. In smaller type toward the bottom of the article the reader learned that Guggenheim also sought to perfect "radio or other aids to navigation and the control for fog flying."[44] Although Guggenheim displayed courage by mentioning fog as a problem yet to be surmounted, the poor quality of navigational instruments made fog, at least in 1926, a seemingly insurmountable problem. Navigation technology of the day was so bad that student pilots were often told by their instructors not to trust their instruments; to compensate, pilots found their way by "contact flying," which meant that the pilot, if he wanted to be safe, had to always fly with the ground below him in sight.[45]

More than anything in 1926, American aviation needed some large doses of positive publicity. Guggenheim had this in mind from the inception of the Fund and was keen to take advantage of the opportunities to dramatize the potential of airplanes, especially through well-publicized tours of commercial airplanes across the country. Previously, Henry Ford's son Edsel had envisioned something similar in sponsoring the Ford "reliability" air tours, which were inaugurated in September 1925. At the same time Ford was promoting reliability, many brave aviators were planning attempts to cross the Atlantic, all of which failed, some of them fatally, until the successful flight of Charles Lindbergh in 1927.[46] Guggenheim was very careful not to associate himself with any transatlantic flights because the risk of gaining terrible publicity for aviation in a failed attempt far outweighed the potential for closely associating the Guggenheim name with spectacular success.

A far safer strategy to follow would be to link the Guggenheim Fund with daring (but not foolhardy) flights only after they had been accomplished. That opportunity came in May 1926 when Commander Richard Byrd, aided by pilot Floyd Bennett, crossed the North Pole in a Fokker trimotor. Immediately after his success, Guggenheim cabled his congratulations in a message that was reported in the *New York Times*: "Your courage in face of tremendous handicaps is an inspiration to all interested in development of commercial aviation. You have demonstrated to the world the increasing dependability of modern aircraft."[47] Guggenheim quickly followed this message with an offer to cosponsor, along with the Department of Commerce, an aviation tour of Byrd's airplane over the entire United States.[48] From October 8 through November 24, Byrd's pilot Floyd Bennett flew over America on a trip that totaled 8,800 miles, introducing the Fokker trimotor to 35 cities with crowds averaging 12,000 in number, which in itself was another notable achievement for aviation.

Certainly, this feat could have stood for many things other than the potential for commercial aviation, as Guggenheim had claimed in his telegram; it could have as easily symbolized the daring of American explorers or the continuing progress of man's domination of the elements. But it was through the well-publicized cheerleading of a legendary explorer and his airplane that

Guggenheim began to carve out his own special niche in America's aviation landscape in the 1920s: he vied to become the interpreter of aviation's accomplishments to the American public. Indeed, the *New York Times* conceded as much by liberally quoting Guggenheim in their article covering the end of Byrd and Bennett's tour. Byrd's accomplishments, according to Guggenheim, were "a remarkable demonstration of the advances already made in aviation that the same plane which carried Commander Byrd and Pilot Bennett over the North Pole is able to complete a swing around the country with no more difficulty than would be found in a motor trip over present day good roads."[49]

Five months later, the Guggenheim Fund found its way into the nation's newspapers again with the official opening of its "safe aircraft competition." On April 29, 1927, Harry Guggenheim spoke before the Yale Club of New York to announce the criteria of the contest, which would require improvements in the stability of airplanes during flight as well as shorter take-off and landing requirements.[50] To date, this was the most visible effort of Guggenheim to inspire wholesale change in the manufacturing of airplanes in America. To effect this change, he pointed to both carrots and sticks. On one hand, the Fund offered not only more than $100,000 to the winner of the competition, but also great publicity on a national stage. On the other hand, Guggenheim lamented a recent tragedy in which two American aviators preparing for a New York to Paris journey perished in their "overloaded trimotor biplane." As one historian has remarked, "Guggenheim referred to the deaths on April 26 of Lt. Comdr. Noel Davis and Lt. Stanton Wooster as a demonstrable justification for the safety competition."[51]

Building on the momentum of this contest initiative, Guggenheim soon afterward refined his argument concerning the relationship between safety and commercial aviation. In "Safety in the Air," Guggenheim made his first appearance as an author in a major American publication: the June 25, 1927 edition of the *Saturday Evening Post*. He repeated an assertion that had been made in previous interviews that aviation accidents drew an inordinate amount of attention from the press.[52] These sensational headlines monopolized the public's attention because of problems inherited from World War I as well as present-day shortcomings of the leaders in the aviation industry. War had doomed the improvement of airplane safety because "the aim of war is to win, and in order to win men must take maximum risks. Safety was necessarily sacrificed to performance. And since war knows no economic law," Guggenheim opined, "the cost of a plane didn't matter, if it could travel faster and maneuver better than a plane built at less expense."[53] Robust commercial aviation, one might then reason, could also be a cause of (rather than just resulting from) better safety standards.

Immediately following the war, according to Guggenheim, American aircraft manufacturers then dodged the obvious problem of flying safety to the detriment of the whole airplane industry:

> Having tabooed the subject of safety, the professional whose aim it is to create a public demand for commercial aviation is caught up in a vicious circle of record making. In the hope of overcoming prejudice against the airplane which results

> from the numerous crashes of flyers, he must set its goal even further beyond the known safety limit and take greater chances. Every commercial manufacturer, designer, and pilot knows that disaster is almost a mathematical certainty under this procedure.[54]

In this very straightforward and public assault on the problem of safety, Guggenheim was obviously positing himself and his Fund as pointing to a very different way to preserve and grow aviation in America. These words would be followed by deeds, Guggenheim concluded: "I express the opinion... after most careful study of aviation in both this country and abroad, that we shall have airplanes that will fulfill the seemingly difficult conditions of the [safe aircraft] competition within the next few years."[55] Although Guggenheim was making a careful and skillful effort to become the leader of American aviation in the spring of 1927, something unexpected happened to force Guggenheim to opt for a strong supporting role instead.

Association with Charles Lindbergh

Charles Lindbergh's now-fabled flight of May 20, 1927 raised the young flier to the level of international superstardom.[56] The transatlantic crossing made his face the indisputable symbol of the promise of American aviation in the 1920s. The *New York Times* followed his every move in Europe and then back on his return trip to America. Eventually, Lindbergh penned his own series of articles for the nation's newspaper of record under the title, "Lindbergh on Flying."[57] Reeling from the adulation he received upon returning to America in June, Lindbergh found some help after visiting with President Coolidge. The president's advisor on aviation, Morrow, contacted Harry Guggenheim to see if he might be able to offer Lindbergh a refuge from the tumult. Harry was, of course, delighted to welcome the aviator to his home on Long Island, *Falaise*.[58]

This meeting with Lindbergh offered Guggenheim a splendid opportunity to begin another aviation tour that certainly dwarfed Byrd's in terms of the interest it generated in the public. As the *Washington Post* reported, Lindbergh's tour, which began on July 20, brought the aviator to 82 cities across America on a 22,000 mile tour of the country with an estimated 30,000,000 people coming to see the *Spirit of St. Louis* up close or in the air. Gratifyingly for Guggenheim, the *Post* also made special note of Lindbergh's remarkable punctuality:

> The hero of the New York-Paris flight of last May, journeying over the country under the auspices of the Guggenheim Fund for the Promotion of Aeronautics, has made a flight as safe and sane as his Atlantic adventure was audacious, and his plane has winged its way from city to city with a punctuality that rivaled the record of the best of fast trains.[59]

The *Post* used Lindbergh's flight as an example of the potential of airplanes to assume their rightful position in the American economy by, the article implies, replacing the train as the primary mode of long-distance transportation in the

Colonel Lindbergh (left) and Harry Guggenheim. (Source: Bettmann/CORBIS).

not-too-distant future, just as Guggenheim himself might have argued. Indeed, the *Post*'s interpretation of the meaning of the Lindbergh flight for the future of American aviation may have been directly influenced by Guggenheim himself.

Although Lindbergh became the face of American aviation in the summer and fall of 1927, Guggenheim attempted to profit from the young aviator's celebrity by interpreting the meaning of Lindbergh (or, more precisely, Lindbergh's transatlantic flight) to the American public. Featured on the cover of July 15th edition of *Forbes*, Guggenheim's article "What Must Be Done to Make Commercial Flying Safe" began with this analysis:

> The millions of Americans and Europeans who first followed the intrepid Lindbergh in their hopes and prayers and then joined in the world's acclaim, are now passing from the personal and heroic element of the imperishable flight to some of the technical aviation problems involved, in a desire to understand them. They are beginning to wonder if the flight is really going to play a part in their own lives. "Do you think I will ever fly to Europe?" is asked of me now with persistence and real interest.[60]

The rest of the article then tries to answer "yes" to the open question of the possibility of flight for the general public by showing that Lindbergh's amazing oceanic crossing was much more difficult than it had to be. To finish a nonstop flight across the Atlantic, Lindbergh eliminated a variety of options that would have made the trip far less dramatic. He could have stopped at Iceland or Greenland on his way, thus lessening the weight of the *Spirit of St. Louis* that was overloaded with fuel and permitting him to include things like a windshield and radio equipment. This may have also opened enough space for an extra pilot who could have relieved him from his arduous 33 consecutive flying hours. In addition, the infrastructure of aviation could be easily improved with current technology, providing pilots with aids in navigation and weather reporting. With this knowledge, the real lesson of Lindbergh's triumph was the probability of making available long-distance flight to the public in the near term.[61] Finally, in publishing this article in America's leading business magazine, Guggenheim effectively created a challenge for himself to realize at least some of his visions for aviation's future by risking his reputation in full view of his peers in the business community. Guggenheim's efforts to associate his Fund with high profile celebrities like Lindbergh and Byrd are indicative of many successful entrepreneurs who rely on external or environmental factors to publicize their ideas or new product offerings.

Guggenheim was so conscious of protecting the forward momentum of the Fund's safety theme that in September 1927 he demanded that the U.S. government place a curb on stunt flying. The *Washington Post* reported: "Guggenheim said he did not believe that all prize offers should be banned but rather be given for the development of better equipment than for spectacular achievement in existing planes."[62] This initiative was not pursued with much vigor by Guggenheim who seems to have uncharacteristically misjudged his ability to put a limit on the conduct of individual Americans who might exercise the right to put their own lives in danger. Fortunately, he soon redirected his attention to endeavors that would, instead, further increase the confidence of future aviation investors and passengers.

Early Support for Passenger Traffic

Indeed, for the next year or so Guggenheim put aside his public-relations pen for the most part and focused on promoting some more tangible improvements that could benefit commercial aviation through a focus on safety and reliability. A major initiative that explicitly combined technical and commercial interests of the Fund was the establishment of a "Model Air Line." The idea for a model airline was actually promoted most vigorously by Harry Guggenheim's father. While Harry may have thought that the Guggenheim Fund was to prepare the way for greater passenger traffic on commercial airlines, Daniel Guggenheim was impatient to do something tangible to create an actual airline industry. In response to his father's demands, Harry Guggenheim successfully summoned top executives from many of the nation's largest airmail carriers to meet him in New York on May 27, 1927.[63] Harry informed them that the Fund would subsidize the purchase of some of the best airplanes available for an airline that would focus on a passenger-only service. In addition, the Fund would promote some of the latest research in aviation-related technologies by supporting one or more passenger-only "routes with radio communications and meteorological services approved by the Department of Commerce's Aeronautics Branch."

Harry Guggenheim was disappointed to learn that all "of the big shots of the air mail in that stage of development threw cold water on the idea of flying passengers," believing that airlines were doomed without a government contract of some sort. The general reluctance of the executives to explore the possibilities of passenger-only airlines were all the more remarkable seeing that their meeting with Guggenheim took place less than a week after Lindbergh's triumph.[64]

One of the two executives who did express interest in the project, Harris M. Hanshue from Western Air Express (WAE), was awarded a generous loan of $180,000 from the Fund on September 15, 1927. WAE was already providing profitable airmail and passenger service between Los Angeles and Salt Lake City. Guggenheim was impressed with Hanshue's record as president of WAE and was happy to find a kindred spirit who believed that the airline industry depended on attracting passengers rather than government airmail contracts for their principal source of revenue. In addition to Hanshue's competent management, the "Model Air Line," whose route would cover the more populous section of the country between San Francisco and Los Angeles, would be supported by state-of-the-art meteorology reports (which also demonstrated the benefits of research spurred by the Fund's university grants) sent to pilots via ground to air radio.

With three reliable 12-passenger Fokker trimotors, WAE began its airline service on May 26, 1928. Despite the wide variety of climate conditions one might encounter between dry Los Angeles and foggy San Francisco, weather reporting stations along the way helped WAE to maintain "99 percent on-schedule reliability during the first seven months of accident-free operation," succeeding in transporting around 3,000 travelers. Although by the time Hanshue paid back the Fund's loan in 1929 the airline had not made a profit (in fact, it had been supported by WAE's more profitable airmail services), he considered that the reliability and safety demonstrated by the Model Air Line experiment had been a success.[65] The *New York Times* agreed, pointing to Guggenheim's support of WAE as an important means to bridge the gap between Europe and America in the development of domestic passenger airlines.[66]

Nine months after the Fund had begun its "Model Air Line," Harry Guggenheim took aim at the most daunting technical challenge facing aviation at the time: blind flight. In a press release on June 22, 1928, the Fund announced that it would "transfer its emphasis from the work of assisting commercial aviation and stimulating public interest in its development, to the consideration of fundamental aeronautical problems," especially "meteorology and the problem of fog-flying."[67] Traveling in fog in the mid-1920s, as Guggenheim quoted Lindbergh in a subsequent publication, was not necessarily dangerous for the experienced pilot. The rule was simple: "Do not attempt to fly through fog. Turn back or land before it is too late."[68] While this could save a life, it would kill any prospect for aviation to be used as a reliable means of transportation. For those pilots foolhardy enough to venture into a fog and trust their instincts alone, accidents and deaths often resulted. To address this problem, Guggenheim announced in August 1928 that the Fund would soon start a "Full Flight Laboratory." This lab was located at Mitchell Field on Long Island. There, navy pilot Lieutenant James Doolittle was put in charge of running the entire project. Doolittle's reputation helped to bring some of the most accomplished people in aviation to Long Island, including Paul Kollsman, an engineer from Germany who made crucial improvements to altimeter technology.[69]

Starting in late 1928, Guggenheim increased his public visibility through speeches as well as publications—from the uplifting, "Giving Wings to the World," published in the popular children's magazine *St. Nicholas*, to "Making Flying Safe," which appeared in one of the last issues of an august journal dating from the early nineteenth century, *Forum*.[70] In the October 1929 edition of the *Harvard Business Review* (HBR), Guggenheim wrote a long piece called "Aviation—Progress in Safety." The recent successes of the Fund had only encouraged Guggenheim's optimism, which moved him to claim in the opening paragraph of the HBR article that his eventual goal was to increase the 7,000 or so private planes in the United States to a number approaching that of private automobiles—which was 24,000,000. "That was the purpose," Guggenheim explained, "behind the international Safe Aircraft Competition," which would provide the public a clear demonstration that "airplanes are inherently no more dangerous than steamships or railroads."[71] After listing the current practical efforts to reduce the risks of airplane stalls, Guggenheim was proud to announce that the Fund had also just overcome the most daunting problem in aviation: "On September 24, at Mitchell Field, Lieutenant James H. Doolittle . . . took off, flying completely blind in a covered cockpit, flew away from the field and returned to a given spot and made a landing."

The lesson to readers of the HBR was clear: manufacturing expertise, business skills, and funding were the only things now needed to make commercial flying a reality. Fulfilling these needs for commercial aviation would require broad private and pubic support, but that was only a matter of will—not of capability:

> In other words, with the commercial manufacture of these instruments, the necessary equipment for fog flying will be neither expensive nor complicated, but of such a nature that it is readily available to the average pilot, and easily

> comprehended. The commercial practicality of the development is, therefore, assured from the start.... The application of [fog flying techniques], however, and the final perfection of the best equipment for all phases of fog flying will require time and effort on the part of commercial and military organizations.[72]

According to Guggenheim, progress in aviation was a reality, not a dream. It would not only "make possible a more intimate contact with our Latin American neighbors with whom our relations have been more limited than they should have been," but it would also open vast economic opportunities for places in the United States and abroad that had lacked access to good railroads or ports.[73]

In the end, Harry Guggenheim's greatest achievement lay in his ability to calm the national mood about aviation and to instill confidence in its future. Guggenheim's entrepreneurship was characterized by a combination of inspiration, visionary goals, and cooperation. He advocated for collaborative working arrangements with leaders in every aspect of aviation—from scientific researchers to business leaders to government regulators. Since Guggenheim was more of a behind the scenes entrepreneur, his tools were cooperative alliances and partnerships between government and business. He was a skilled public-relations specialist who selectively used press releases and media coverage to showcase advances in aviation in an effort to galvanize public and private support. He essentially became *the* ambassador for flight.

Having succeeded in laying the technological and cultural foundation for safe and reliable passenger flight by 1930, Guggenheim took the next logical step in his plans. By assuming the role of President Hoover's ambassador to Cuba, Guggenheim worked to realize the promise of international peace and prosperity that he believed would be ushered in by the age of aviation. Cuba would be the lynchpin in an aviation infrastructure that would greatly increase economic and social ties between North and South America. In an article written for the *New York Times* in September 1929 (one month before the Great Crash), Guggenheim shared this vision in an intimate tone with his readers: "More important than the economic relations of the Americas are our social relations. If you believe with me in President Hoover's dictum in his address at Lima, Peru on his recent visit to South America: 'It is a benevolent paradox that to destroy the distance between peoples is to construct friendship between them,' then indeed the airplane will be twice blessed."[74]

Uplifting though this vision was, not everybody reading Guggenheim was convinced that the airplane would usher in perpetual fair weather for the world's future. In a *New York Times* review of *The Seven Skies*, a collection of selected writings by Guggenheim published in 1930, the aviator Captain T. J. C. Martyn questioned, for example, Harry's optimistic assessment of the Fund's progress in addressing the problem of flying through fog: "It is true that a new principle [of blind flight] was born, but it took one of the world's most expert pilots to demonstrate it.... The hard fact remains that the conquest of fog has just begun, for all practical purposes, and the fight may yet prove to be a long one."[75] More recently, one of Guggenheim's biographers has concluded that the many financial difficulties faced by airlines in the United States might have been mitigated had Harry Guggenheim not insisted on eventually eliminating all subsidies for aviation.[76] Yet despite these blind spots, as Martyn concluded in 1930, Guggenheim's overall record was astonishing:

> There are few men who have done more to aid in the solving of some of the current problems of aviation than Mr. Guggenheim, and if he lets his optimism run away with him now and then, it is to be remembered that he has been a close witness of some remarkable achievements, so remarkable, indeed, as to make any man believe that we are nearer an aerial Utopia than we really may be.[77]

CHAPTER 2

Juan Trippe's Early Entrepreneurial Efforts

It was October 18, 1927, and Juan "Terry" Trippe, the new president of Pan American Airways, was just days from becoming a failure in the nascent airline industry. Trippe had already begun two airline projects that had ended almost as quickly as they had begun. Now, embarking on his third, Trippe ran into a pressing problem. Pan Am's head pilot, Ed Musick, was supposed to be carrying the mail from Key West, Florida to Havana, Cuba to fulfill a mail contract granted to Pan Am from the U.S. Post Office. Unfortunately, Musick could not get to the Keys; he was stuck in Miami with a new Fokker airplane because the runway at Key West's airport had been reduced to mud by some recent rains.[1]

Although at this moment Pan Am was effectively an airline without an airplane, Trippe still possessed something truly valuable: exclusive American landing rights to Cuba granted by the Cuban government. These rights could be very lucrative for an airline carrying mail between the United States and Latin America. However, the post office mail contract, which would help to realize the monetary potential of those landing rights, would be withdrawn if Pan Am did not fulfill its promise to deliver airmail to Havana from the United States by October 19, 1927.

How did Trippe find himself in this tight situation (and others like it), and how did he manage to survive and prosper? Although Trippe certainly learned from some early mistakes, he never relinquished his monumental ambition. From a very young age, Trippe understood aviation's potential to interconnect the world through overseas air routes, and he was very impatient to realize those lofty dreams.[2] To the chagrin of many of his colleagues, Trippe seldom put any effort into persuading his business partners to agree to his goals. Instead, he often forged ahead with his own ideas. Luckily for Trippe, his visions of the airline industry along with his tenacity aligned with United States' foreign policy in the late 1920s and early 1930s to create a powerful American airline of impressive (perhaps almost unimaginable at the time) international scope. In many ways, Trippe displayed the classic entrepreneurial characteristic of hubris, which enabled him to forge ahead despite tremendous obstacles. This deep, unyielding personal belief in his vision for a global airline propelled him to succeed for decades, yet, as we will see, it ultimately became part of the downfall for Pan Am. The story of how Trippe learned from his early errors to create and then to dominate the international air travel market from the United States is a complicated one that demonstrates not only how he tried to shape the context of his times, but also how he significantly influenced the field of international aviation itself.[3]

Trippe's Early Years

Trippe was born into an upper-middle-class family in Seabright, New Jersey, and he came of age during the barnstorming, stunt-driven time of the early aviation industry. When he was 10 years old in 1909, his father brought him to an air race over Long Island where Wilbur Wright flew around the Statue of Liberty to the amazement of thousands of spectators.[4] Once he experienced air flight for the first time in college, his fate was essentially sealed.

Trippe's father was an investment banker and broker in New York City who expected his son to follow him in the banking business. What he lacked in natural intellectual abilities, Trippe made up for in steadfast industriousness and determination. He also developed a strong ability to debate his viewpoints. He attended high school in Pottsdown, Pennsylvania, and graduated from the Sheffield Scientific School at Yale University. Trippe's time at Yale was interrupted twice. The first time occurred when he served in the Naval Air Corps during World War I. Though the Armistice was signed before Trippe saw any action, the experience further reinforced his love of aviation. The second time he briefly left Yale was when his father passed away. Trippe helped to settle his father's estate and then returned to finish his studies.[5]

In 1920, Trippe helped to form the Yale Aeronautical Society, which was made up of 50 Yale students and pilots who had flown for the military during World War I. In that same year, Trippe represented Yale in a race sponsored by the "Intercollegiate Flying Association," which was made up of about a dozen "air clubs" from Ivy League colleges.[6] The course was a 25-mile-long, four-cornered route on Long Island. Trippe tried his best to get an edge by "modifying the incidence of the dihedral," which meant actually changing the angle at which the airplane wing connected with the plane's fuselage. Trippe recalled skimming the treetops on his way to the finish line and then winning the race by a few seconds.[7] The friends who joined him in the Flying Club and on Yale's varsity athletic fields were not just run-of-the-mill college buddies. They were influential, wealthy, and well-connected friends whose last names were Whitney, Rockefeller, and Vanderbilt.

As a tribute to his father and in an effort to help support his mother, Trippe, upon graduating from Yale, spent two years as a bond salesman for the Lee, Higginson and Company investment firm. He called them "the dullest years of my life," though he continued to solidify relationships with influential and wealthy colleagues. In 1923, he left Wall Street to follow his passion for aviation.[8]

Long Island Airways

As of 1923, there was nothing like an "airline industry" that Trippe could model a business after. There was hardly even an airplane manufacturing industry because the glut of war surplus planes made airplanes cheap and easily available.[9] His first airline, Long Island Airways, catered primarily to the rich. Trippe's airline found a potential competitive advantage in a seacoast location where slow planes could still outpace boats and even railroads that had to follow the contours of the land. In many ways, Long Island Airways was modeled after the St. Petersburg-Tampa Airboat Line.

Although the St. Petersburg to Tampa air route may seem to be an unlikely and underpopulated place to start an airline, there were two factors that made this route a good place for the launch of the first experiment in scheduled airline service. First, airplanes' cargo and passenger capacity had not progressed much since the Reims aviation meet of 1909; therefore, one did not need to serve large population centers to fill an airplane (the sea plane used by the St. Petersburg-Tampa Airline could only hold one pilot and one passenger). Second, slow airplanes could still not compete head-to-head with relatively speedy trains. The only clear advantages airplanes held over railroads was the ability to travel over water and to other places where rail or road infrastructure were not in place.[10] The waterway between St. Petersburg and Tampa was tailor-made for this business. The wealthy summer vacation crowd provided a built-in customer base. Although the St. Petersburg-Tampa Airboat Line folded in April 1914 after only four months of service (when many of its target customers returned to their northern homes), it had successfully transported 1,200 passengers.

After investing $1,500 of his own money, Trippe gathered another $3,500 (at $5 a share) from wealthy friends and relatives. With that money, Trippe showed some uncharacteristic patience in purchasing the aircraft for his airline. When Trippe learned of an auction of Navy Training Planes, he bid on seven Aeromarine 49-Bs. After the U.S. Navy rejected all the offers for their planes as being too low, Trippe returned with the exact same offer when bids were taken again. This time, he got the planes for the price he wanted.[11]

During the summer, Trippe arranged to transport wealthy socialites to their vacation houses in and around Long Island. He looked a little farther away to the shores of New Jersey and made some deals that included round-trip packages with Atlantic City hotels. Although these routes attracted paying customers, the two-seater (including the pilot) Aeromarine could be made even more profitable with an extra passenger. This could be done by substituting the original 90 horsepower engine with a 220-horsepower French-made "Hispano-Suiza" engine. Although putting in the new engine could only be done by further modifications (including the use of a smaller propeller) and an extra passenger could only be squeezed in by moving the gas tanks from inside to outside the fuselage, Trippe succeeded in making the adjustments. A two-seater could attract a new clientele—couples—and promise to double the revenues per flight.[12]

Trippe was certainly a master of technical innovations for airplanes, but it was still difficult for Long Island Airways to make a consistent profit. The need to find innovative ways to bring in cash for the services that could be rendered by an airplane inspired Trippe to look beyond the borders of the United States for new clients. Trippe again leveraged his connections with wealthy Yale classmates for letters of introduction to huge businesses like United Fruit, a company deeply involved in the production and distribution of Central American produce since the nineteenth century. Trippe thought that his airplanes might help expedite aspects of United Fruit's business, much of which had to negotiate mountainous terrain, such as the scenic but treacherous roads of Honduras. When he learned that United Fruit's ships had to obtain official stamps at Honduras's capital to finalize transactions that were taking place at the coast, Trippe suggested that his Aeromarine planes could cut travel time between the two points from three days to a few hours. United Fruit agreed and used its considerable influence to acquire landing rights for Trippe's airline.

Trippe also looked north where he offered the services of two of his planes to logging companies located in remote areas of the Canadian forest.[13] He reasoned that these areas were the most difficult to traverse, and although air flight was slow in its early days, it was far superior to other transportation options in sparsely inhabited areas. With this strategy came increased danger and uncertainty (often no visibility, unsafe runways, and unpredictable and uncontrollable weather conditions), but it was an essentially untapped opportunity. Although most of Long Island Airways' planes that ventured far from their home base were eventually wrecked in an accident of some sort (demonstrating an early instance of Trippe stretching his company beyond its current capabilities), these ventures abroad provided important lessons for the young airline executive.

While Long Island Airways made the most of their unique international offerings, its domestic beach flying business suffered from intense competition, and Trippe was unable to develop a long-term, sustainable revenue stream from his initial ventures. After 18 months, he sold Long Island Airways and began looking for a more viable airline opportunity.[14] In an interview in 1976, he claimed that his perspective on the airline industry had developed considerably by 1924. He recounted that an

> airline must have a route.... It must have regular schedules. It must carry mail. The future of the airline business was in a substantial company.... The Post Office certainly would never award an airmail contract to an airline like Long Island, capitalized at only $5,000, and a postal subsidy was essential if an airline was to survive long enough for passenger and freight traffic to build up.[15]

Given the state of the airline industry in the mid-1920s, these ideas made sound business sense. After World War I, the U.S. government sought to improve the speed of mail delivery through the development of a transcontinental airmail route.

The initial government experiment began with an appropriation of $100,000 and the request for bids for the construction of five postal airplanes.[16] With the development of a series of short run flights between major U.S. cities, the Post Office Department demonstrated the viability of airmail and secured an additional $1.5 million in government appropriations in 1921 to improve landing strips and add lighting for night flying. Over the next four years, the Post Office Department continued to build a nationwide airmail system, which helped generate momentum for further government subsidies and more airline carriers.[17] Early enthusiasts of airmail included bankers who looked to airplanes as a useful means to speed up the transfer of checks and other financial documents. By 1923, the Post Office's airmail service had flown more than 10 million miles and delivered more than 67 million letters.[18]

The 1925 passage of the Contract Air Mail Act, also called the Kelly Act for its chief sponsor, Representative Clyde Kelly of Pennsylvania, essentially turned the entire job of flying the mail over to private carriers.[19] With this Act and the subsequent Air Commerce Act of 1926, the government awarded airmail contracts to private companies and provided funds for building and maintaining the safety of the nation's airports and airways.[20] The post office received more than 5,000 applications for the first 12 air routes that it awarded in 1925.[21]

Colonial Air Transport

Now that the government was committed to funding private airlines to transport federal mail, Trippe moved aggressively to take advantage of this new context. To increase his odds for success, Trippe incorporated more than one airline; among them were Alaska Air Transport and Eastern Air Transport. Running an airline in Alaska promised the same advantages as flying in Honduras: rough terrain where an airplane could easily beat any of the land- or sea-based transportation. Unfortunately, Trippe was outmaneuvered by a formidable foe: an Alaskan dog sled. Men driving dog sleds were the traditional mail carriers of Alaska. They beat Trippe to Washington, D.C. and argued that Alaska's "Star Route mail contracts" made no provisions for airplanes. Although Trippe took his case to the U.S. Post Office and then pushed for new legislation to be written in Congress, his petitions were denied. This would be one of the last times Trippe would ever fail so decisively in his lobbying efforts with the federal government.[22] With that lesson learned, he went on to another project.[23]

Trippe incorporated Eastern Air Transport in Delaware in September with the hopes of winning a mail route. He prepared this bid well. As early as April 1925, Trippe conducted a marketing survey to assess the potential profitability of air service between New York and Boston. Businessmen using express mail would provide their primary source of income. Trippe planned connections down the East Coast all the way to Havana that would be added to this original route. He then gathered the estimated start-up money that would be necessary for this new airline—the significant amount of $250,000—from wealthy aviation enthusiasts who would become part of the company's board of directors.[24]

Unlike the case with his failed Alaska Air Transport, Trippe went to Washington, D.C. soon after incorporating Eastern Air to lobby for carrying the mail on what was known as Contract Airmail Route #1 or CAM 1. Upon arriving in Washington, D.C., Trippe's plans hit a snag: a rival group was bidding on the same airmail route. This second company was called Colonial Airlines, and it was run by people with sounder business backgrounds and higher profiles than the 26-year-old Trippe, the most prominent being John H. Trumbull, who was also the Governor of Connecticut. The post office urged Trippe's Eastern Air Transport and Colonial to merge, a move that would probably have doomed executives less ambitious than Trippe to second-tier status after the merger. Instead of fading into the background, he called again on his Yale friends from the Vanderbilt and Rockefeller families, among others, to join in his venture. They came, and with their money Trippe gained some important leverage. He used it in the merger to become one of four vice presidents of the new Colonial Air Transport. The company soon won the rights to provide airmail service on CAM 1.[25] Two stipulations of the contract were particularly important: (1) the airmail service was to begin by July 1, 1926, and (2) the airmail was to be transported by single-engine planes.[26]

In Colonial Air Transport, Trippe acted very much as he had when he promoted Long Island Airways. He was headstrong as well as indefatigable and involved himself in every aspect of the business—from hiring pilots to buying planes for the infant corporation, to persuading people to buy Colonial's stock with the promise that the airline would eventually take over the Key West-Havana route from the recently defunct Aeromarine Airways.[27] There was, unfortunately for Trippe, one significant difference: he was no longer the owner and president of his own business. Nevertheless, the original chairman of the board, Governor Trumbull, was soon displaced by a large stockholder whom Trippe had convinced to join Colonial, Theodore Weicker, the wealthy father of another Yale classmate. The next year, Trippe added to his power at Colonial by convincing the directors to hire a president, John F. O'Ryan, who had some experience in transportation and who (Trippe believed) could be easily manipulated.[28]

Besides his natural desire to run things on his own, Trippe tried to amass this power within Colonial because the executives from the Trumbull bloc of the airline did not share Trippe's visionary enthusiasm. These were conservative investors from New England who were not tightly enmeshed with Trippe's Wall Street crowd and, more importantly, had no experience in aviation. Their goals were simple: start making money as soon as possible from transporting the mail over the important commercial route between Boston and New York. Trippe saw things differently: he wanted to carry passengers domestically *and* internationally, too.

Trippe found a way to combine his goals of adding passengers to his airline and flying abroad to Cuba: promoting the airplanes of a new friend in aviation, Tony Fokker. Fokker was a truly transnational product of the aviation age. Having started as a manufacturer for the German military during World War I, he then moved to the Netherlands before the victorious Allies destroyed his factories in Germany. In 1920, he developed the ground-breaking "F-4" that he constructed for the American Army Service. Besides being able to carry 11 passengers, the F-4 broke many speed, distance, and endurance records.[29] Trippe (ever appreciative of technological advancements in aviation) and Fokker must have met soon after Fokker established the Atlantic Aircraft Corporation in 1924. The company was located in Teterboro, New Jersey, less than 20 miles from Manhattan.[30]

Trippe's assessment of the aerodynamic viability of Fokker's aircraft was spot on, for Fokker, along with the Ford Motor Company, was creating some the best airliners of the day.[31] The airplane that caught Trippe's attention was the F-7A. In these early days of flight when airplane engine failure was one of the main causes of accidents, Fokker alleviated this problem by replacing one large engine with three smaller ones. He was able to do this by taking advantage of a recent development made by the Wright Aeronautical Company: the Whirlwind, a powerful engine that used air instead of heavier liquids to cool itself. This was the very engine used by Charles Lindbergh in his *Spirit of St. Louis* to traverse the Atlantic Ocean two years later.[32]

In need of cash at this time, Fokker was happy to do anything he could to promote his airplane. Trippe hired a press agent to arrange some highly publicized flights of the F-7A over New York City, billing the plane (that Trippe had decorated with a Colonial Air logo) as a "flying Pullman car" where passengers could comfortably enjoy urbane pleasures such as a cup of tea and a functioning toilet. In December, Fokker and Trippe joined together in a "giant survey trip" of Florida, stopping frequently to allow the airplane manufacturer to deliver speeches about the F-7's virtues.[33]

More importantly for the future of the airline career of Trippe, the survey trip of Florida was extended to Cuba on Christmas day. Trippe had long anticipated this journey and acted quickly once he arrived. The Cuban president, Gerardo Machado, was impressed with Trippe and Fokker who demonstrated that the F-7A could fly with two engines and then with one. The next day Trippe found a lawyer in Havana who had been recommended to him by a Yale alumnus. The lawyer prepared an agreement that granted Trippe personal and exclusive landing rights in Cuba. Machado signed the documents, giving Trippe a foothold in the Caribbean in 1925.[34] Now, the challenge would be to persuade the rest of Colonial Air Transport to follow his lead.

Many of Trippe's colleagues at Colonial Air were not impressed. Looking for markets down the East Coast was all well and good, but where were the airplanes to fly CAM 1 (the Boston to New York airmail route)? The core business was being neglected as Trippe pursued his larger ambitions. Trippe's grand design had inspired him to order some of the best and most expensive planes around: two F-7As as well as two Ford trimotors. (The Fokker airplanes cost $37,500 each.)[35] Waiting for these machines to be constructed cost precious time. Adding insult to injury, there were plenty of other less-sophisticated, single-engine planes that were perfectly adequate to transport mail and were in compliance with the airmail contract that Colonial had signed with the federal government. Owing to Trippe's grand plans, as of March 1926, Colonial was generating no revenues—but it was receiving bills from Fokker's Atlantic Aircraft Corporation.[36]

Even with these poor returns and no trimotor airplanes delivered (Colonial had been forced to use the single-engine Fokker Universals), Trippe went forward with plans to continue to expand the airmail routes for Colonial Air. In January 1927, despite the strong objections of many board members, Trippe submitted a bid for the New York to Chicago mail delivery route. Many of

Trippe's colleagues on Colonial's board, especially Governor Trumbull, were incensed.[37] Hoping to slow down Trippe's willful ways, Trumbull and O'Ryan (no longer as tractable as Trippe had hoped) put the matter before the stockholders. Although Trippe lost the vote by a small margin, his power as a "behind-the-scenes" director of the airline had been severely compromised. Trippe resigned, bitterly condemning those who had opposed him as unable to understand his vision of creating the largest domestic airline company in

Three pilots for Colonial Air Transport, Inc. are congratulated by Managing Director Juan T. Trippe following their successful first airmail trip between Boston and New York. (Source: Bettmann/ CORBIS).

America. But as of May 1927, vision was just about all Trippe had left to show for his efforts at Colonial Air Transport.[38]

Trippe exhibits some of the classic tenants of an entrepreneur, a willingness to continue in the face of adversity. A characteristic of many entrepreneurs in start-up industries, which we see in today's high technology world of Silicon Valley, is the ability to fail and then get up and try again, and sometimes again. Successful entrepreneurs tend to possess a level of self-confidence and resilience that enables them to confront and conquer challenges. As Trippe experienced, success often does not come right away. In addition to resiliency, successful entrepreneurs are often great learners. In fact, it is their ability to learn from their mistakes that gives them the confidence and resilience to venture forward.

Third Try—Aviation Corporation of America

A month after leaving Colonial Air Transport, Trippe revived his own fortunes again and helped to found yet another corporation dedicated to air travel: Aviation Corporation of America.[39] This time, Trippe was able to create a company that was much friendlier to his interests. Yet again, he drew on some Yale buddies with deep pockets, some of whom, like Cornelius Vanderbilt Whitney, were adventurous pilots who shared his passion about flying.[40] Eventually, Trippe also brought along some impressive outside investors such as W. Averell Harriman (son of railroad tycoon E. H. Harriman and a future prominent politician and diplomat) and William Beckers (the founder of a large company that would later be acquired by Allied Chemical).[41] Trippe's persuasive gifts were given a fortuitous boost at the time by Lindbergh's successful crossing of the Atlantic. As we saw in the previous chapter, Lindbergh's solo flight fueled financial speculation across the entire airline industry. Trippe could even boast to potential investors of having attended Lindbergh's take-off from Roosevelt Field on Long Island.[42] He took full advantage of this opportunity and managed to raise more than $250,000 for his new venture through the sale of stock.[43] Trippe's biographers explain that stock was used exclusively to finance the company as bankers were not at all enthusiastic about the prospects of Trippe's company. Because money from private investors was flowing very easily into start-ups like Aviation Corporation of America, many of the companies that sold stock to the public actually had no assets. Instead, they were often formed simply to bid on potential airmail routes. And, when "several promoters converged on the same route, two or more joined forces and ganged up on their competitors."[44]

Trippe and his associates had big plans for Aviation Corporation of America. They wanted to start with developing three different regional routes simultaneously: "a service out of Buffalo, another to the society resorts within a radius of New York and Baltimore, and a network of routes to the Caribbean and South America."[45] When Aviation Corporation of America began on June 2, 1927, manufacturers were making steady strides forward in improving airplane performance. By 1928, Trippe's old friend Tony Fokker had already come out with the F-10, which could carry 12 passengers a distance of 700 miles.[46] But the company's directors saw far beyond the limitations of the commercial aircraft of the day. They discussed the real possibility of crossing the Atlantic as well as the Pacific, which, at 5,000 miles across, exceeded the F-10s range by more than 600 percent. Obviously, they were confident that technology would develop quickly enough for them to realize their dreams *and* make a profit. They also contemplated the seemingly impractical task of completing intercontinental travel over the Arctic Circle. They were prescient: Wiley Post grabbed the nation's headlines after traveling around the globe around the fringes of the Arctic in 1931.[47] In later years, Whitney summarized the reasons behind many aviators' buoyant attitudes in the 1920s: "We were flying since we were eighteen; we had great faith."[48]

Trippe's daring plans for air routes that spanned the globe were matched by his fearless pursuit of Lindbergh. Not even a month after Lindbergh's triumphal arrival in Europe, Trippe managed to prepare the ground for associating the young transcontinental superstar with his own international ambitions. Although Lindbergh was physically daring and possessed a steely will, he was socially withdrawn. Only 28 years old, he was overwhelmed by his sudden fame as well as all the lucrative offers from dozens of parties who dreamed of the marketing windfall of having Lindbergh on their side. Although Trippe could match any of the aggressive tactics employed by many of Lindbergh's corporate suitors, he sensed that a softer approach might be more effective.[49] During Lindbergh's stay in New York in June 1927, Trippe managed to arrange a rare 15-minute audience with the new prince of the skies.[50]

During that interview, Lindbergh seemed to see Trippe as a kindred spirit; he confessed to Trippe that he was baffled about what to do with the myriad offers coming to him every day. Acting as a good counselor, Trippe acknowledged Lindbergh's difficulties and suggested that he wait a week before deciding upon anything. Trippe attempted to assuage his fears further with an astute suggestion: although he also had an offer to make, he would not negotiate directly with Lindbergh until the pilot hired a lawyer. Lindbergh's forte was aerial adventure, not business. Years later, Trippe recalled this gambit as having cemented their relationship: "that impressed Lindbergh more than anything else."[51] True to his word, Trippe waited a week before making another appointment with Lindbergh. Although he still did not have a functioning airline by late June, Trippe could vividly describe an aviation company that would connect the United States with the Caribbean, South America, and even more far-flung destinations around the globe. They agreed in principle that Lindbergh would be associated with Trippe's airline in some public capacity. In the meantime, Lindbergh made his triumphal tour of the United States sponsored by the Guggenheim Fund while Trippe acquired some airplanes to fly.[52] Carrying the exclusive landing rights to deliver mail to Cuba, Trippe and his colleagues aimed their efforts first to acquire the U.S. airmail contract to service the Key West to Havana route (called Foreign Air Mail route 4 or FAM 4).[53]

Competitive International Landscape

Although aviation was in its infancy at the time when Aviation Corporation of America was planning to transport mail and passengers between the United

States and Cuba, a rich international history of air rivalries, regulations, and treaties had created a propitious moment for an American company interested in connecting the northern and southern halves of the Western Hemisphere. The development of airplanes in the early 1900s, along with improvements in related technologies—such as the lighter-than-air dirigibles developed with particular skill in Germany—provided the technological capstone to an age of internationalism that had been kicked off by the development of reliable transoceanic steamship travel in 1865.[54] The inauguration of the modern Olympic Games in 1896, led by Frenchman Baron Pierre de Coubertin, for instance, could have only been made possible by the ever-increasing economic and cultural contacts between people from all over the globe that depended on improvements in transportation and communication. A mercantile system that connected the world through global trade certainly predates all these technological developments; what these technologies seem to have done is to enable the traffic of large numbers of people who no longer had to be maritime adventurers to travel.

There was certainly much to celebrate in the dramatic "shrinking" of the world through various means of modern transportation. But there were other implications stemming from the growth of air travel that made many nervous. With airplanes and dirigibles, ancient geographic boundaries—say, the Alps or the English Channel—could soon be surmounted almost effortlessly. On one hand, flight had the potential to increase commerce between nations; on the other hand, it contained the possibility of being exploited as a very innovative military weapon.

Great Britain, who had historically felt secure because of its detachment from the European continent, suddenly felt much more vulnerable as flight technology progressed. In 1911, parliament enacted the "British Aerial Navigation Act, which maintained the air above Britain, her Empire, and her Dominions, was sovereign and inviolable."[55] This law became a major impediment for German aspirations to exploit the potential for its dirigibles to take the leading role in aerial commerce in Europe as well as in the European colonies. Like Great Britain, the United States long valued its separation from European powers by a large body of water—in this case the Atlantic Ocean. To the surprise of many, that natural barrier would also become vulnerable to powered flight in just a few years.[56]

World War I spurred great improvement in aviation around the world that the United States could no longer ignore. The Assistant Secretary of War Benedict Crowell sounded this warning in a report to his superiors in 1919:

> The development of aviation is progressing so rapidly at this time that it is difficult even for those in close touch with it to keep up with its progress. During the past two months the Atlantic has been crossed four times by aircraft; first by a seaplane of the American Navy, second, by an airplane of Great Britain, and finally, by an airship of Great Britain which has twice demonstrated its ability to fly between England and America. All of this has been accomplished without the loss of a single life.[57]

Crowell, writing as the chairman of the American Aviation Commission whose mission was to assess European aviation immediately after World War I, also noted that countries such as France, England, and Italy were making great strides in aviation thanks to each government's willingness to fund private aeronautical companies with public monies. "America," he concluded, "has lost its [i.e., aviation's] development to other nations, and too late, realized the mistake of this neglect." Even more alarming, Germany was actually making plans to fly a Zeppelin across the Atlantic to the United States.[58]

As in 1910, the United States did not sign on to an international aerial treaty. But in nine short years, its indifference to the state of European aerial technology had been replaced by alarm. Knowing that it would take some years for American aircraft manufacturers to catch up to Europe, the American government employed some strategies to bide for time. First, the United States did not want any single European country to become dominant in aviation. To increase competition between European countries in the aftermath of World War I, the United States bargained hard with its allies to make sure that Germany was allowed to continue to develop commercial (rather than military) aircraft.[59] Second, in refusing to become a signatory to the Convention Relating to the Regulation of Air Navigation (which was signed by 26 countries in Paris in 1919), the United States would not have to abide by an agreement to "accord freedom of innocent passage above its territory to the aircraft of other contracting States."[60]

The United States was particularly wary of how airplanes might upset its interests in Central and South America. Since the declaration of the Monroe Doctrine in 1823, the United States had become accustomed to having its own way in the Western Hemisphere. But with the combination of airplanes that could steer clear of any navy and the 1919 Treaty that allowed for "freedom of innocent passage" of aircraft from participating nations, the stage could be set for a reassertion of European power in America's back yard. In 1920, an internal report in the U.S. Department of War painted just such a scenario: "in case some Mexican Government should become a signatory to the Aviation Convention . . . German planes [could fly] over Mexican territory and thus near the American boundary. It is conceivable that such a right of flight might be exercised by Germans in a way highly distasteful to the United States."[61] In less than a year, this prediction was beginning to come true.

In fact, some unexpected developments farther south became far more worrisome for many in the U.S. government. In South America, Columbians of German origin began an airline called SCADTA (*Sociedad Colombo-Alemana de Transportes Aeroes*, or "Colombian-German Company of Air Transport") in 1919. This company would soon lay claim to the title of the oldest successful airline in the Western Hemisphere. The company planned to use seaplanes to fly over Colombia's steep mountainous terrain and employ the country's many rivers as landing strips. Although the idea was promising, the company was cash-strapped in its first years. A citizen of Austria who was disillusioned with the state of his home country after World War I, Peter Paul von Bauer, learned about the airline and decided to rescue it by investing his own money in the venture in 1922. Immediately afterward, with SCADTA's service improving, Colombians embraced the foreign-owned airline as a sign of the growing strength and modernization of their own country. With the full support of the

Columbian government, von Bauer launched a scheme in early 1925 to provide mail and then passenger service to Central America, Cuba, and Key West. This plan was made all the more plausible because the American airline that had been attempting to establish routes between Florida, Cuba, and Bahamas with plans to fly to South America—Aeromarine Airways—had folded in 1924. Most significantly, the first stop on von Bauer's ambitious venture would be the Panama Canal Zone.[62]

Traveling to Washington, D.C. to lobby his case to the U.S. government, von Bauer received mixed signals from different departments of the government. While the Commerce Department was interested in the business possibilities that could be opened via SCADTA, the War Department was particularly opposed to granting any foreign airline access to the Canal. Von Bauer was a very charming and well-spoken lobbyist for his airline and made his case to sympathetic ears at the post office and the Commerce Department. But he beat a hasty retreat back to Columbia after the unveiling of a competing airline initiative by Major Henry H. "Hap" Arnold (an early aviation advocate who later became the commander of the U.S. Army Air Force during World War II). Arnold brought a plan to Postmaster General Harry S. New around April 1925 to create a new airline that would offer an American alternative to what Arnold perceived to be a German threat in the Caribbean. The name for the new airline was Pan American Airways. With Arnold's close ties to the government, Pan American soon won the Key West to Havana airmail contract from the post office on July 19, which could be revoked if regular mail service did not begin within three months.[63]

The incorporation of Pan American Airways resulted, in part, from the rapidly improving field of aviation that Trippe and his associates at Aviation Corporation of America had already envisioned for many years. Yet whereas "Hap" Arnold created Pan Am as a company that could fend off a German airline "offensive" from south to north (a strategy that was endorsed by President Calvin Coolidge's administration), Trippe had always approached Cuba as a launching pad to create an aviation empire that would push aggressively from north to south.[64] After the persistent lobbying of von Bauer—along with warnings from some of Germany's wartime foes concerning SCADTA—many in Washington, D.C. were beginning to see things Trippe's way.[65] The time was finally perfect for Trippe to begin to act on the world stage to realize his long-held ambitions.

James H. Smith, Jr., a Pan American executive in the 1940s, described the powerful air of confidence Trippe conveyed to the people he worked with: "Trippe visualized success.... It never crossed his mind that if we were going to open a route that it just wouldn't go off like clockwork. He didn't worry. It wasn't a problem—not for him, not for anyone. He figured, 'I'll get it set up and tell the guys to do it.' "[66] In addition to this iron determination, Trippe was also flexible—practical enough to know that each new problem he met might require a novel solution.

Although Cuba promised to open the door to Latin America for Trippe, it was also attractive to other airlines because it offered a friendly port for Americans who wanted to escape Prohibition (1920–1933). Armed with his landing rights to Havana, Trippe worked on consolidating two rivals to the Key West-Havana airmail route: "Hap" Arnold's Pan American and another rival airline named Atlantic, Gulf and Caribbean, directed by the formidable Wall Street banker and chairman of the Curtiss-Wright airplane corporation, Richard Hoyt.[67] After learning that Trippe held the exclusive landing rights to Cuba, the 40-year-old Hoyt approached Trippe (then 28 years old) and expected to create a favorable deal for himself. Instead, Trippe held his ground and agreed to make Hoyt chairman of the board of their merged airline company while giving Trippe and his investors enough shares to control the corporation.[68] With these two parties in alignment, Hoyt and Trippe approached Arnold's group, headed by former U.S. Navy pilot John Montgomery. Montgomery was confident that Pan American did not need to negotiate because they had a head start: an inside track to the U.S. airmail contract, an airfield in Key West, and ongoing negotiations with the Cuban government. After reacting with disdain to the claim that Pan Am could not proceed without Trippe's landing rights, Hoyt invited Montgomery to take a ride on his yacht to visit the Assistant Postmaster General W. Irving Glover who was in Florida at the time. Glover confirmed: the mail route would not be formally authorized unless the three companies worked together.[69]

Executives from Pan American reluctantly agreed to work with Hoyt in an effort to salvage their newly won foreign airmail contract. Under a new arrangement, Pan American was purchased by a holding corporation controlled by Hoyt. Almost immediately after this transaction, 52 percent of its shares were then purchased by Aviation Corporation—behind the backs of Arnold and Montgomery. Pan American then became a subsidiary of Aviation Corporation of America, and Trippe was named its president and general manager on October 13, 1927.[70] That same day, Trippe announced that Pan American was planning to serve the east and west coasts of South America as far down as Valparaiso, Chile.[71] But before that could happen, Pan American would have to fly to Cuba within the next six days.[72]

A few days later, a Fokker aircraft that the Aviation Corporation of America had ordered months before arrived in Miami, but it was useless on October 18th because of the muddy conditions at Key West's airport. With Pan American employees making inquiries on the telephone from New York and Key West, an amphibious Fairchild FC-2 undergoing repairs in Miami (owned by another small airline) was found. Pan American managed to charter the airplane for $175. The next day, 30,000 letters were delivered to Havana, and Trippe's company formally secured the FAM 4 contract. Pan American Airways—Trippe's third airline in three years—had officially begun.[73]

Although Pan Am's history can be traced back to one specific time and place, the successful trip to Havana acted like a big bang, releasing the energy to expand the airline's business as far south as it could go. Trippe and his associates worked on every front in the northern and southern hemispheres—in finances, negotiating and influencing U.S. government regulations and subsidies, acquiring aircraft, marketing, and negotiating with foreign governments (in public and in secret).

Trippe was particularly adept at three important matters that were of vital importance to securing Pan Am's fortunes in the immediate future. First, and perhaps most importantly, Trippe was an insider's insider in the halls of the U.S. government. He found very loyal allies to help him promote the idea of the United States using Pan American as a "chosen instrument" of U.S. foreign policy, helping to promote U.S. interests abroad by means of aviation. Second, by bringing on Lindbergh as an employee of Pan American, Trippe had a powerful marketing tool without equivalent in the international airline market. Finally, Trippe successfully overcame the diverse obstacles presented by domestic and foreign competitors in his bid to be the sole U.S. operator of international flights and to construct a hemispheric airline empire.

Trippe's experiences with his two previous airline ventures had taught him that an international airline based in the United States had little chance of survival on competitive mail contracts alone. The initial investments in surveying Central and South America coupled with the needed infrastructure to support an airline so far away from the developed United States would be quite substantial. Those expenses would be coupled with the extra costs of acquiring aircraft big and sturdy enough to carry mail and passengers for long transoceanic journeys. Trippe believed that the government would have to award more lucrative contracts for longer periods of time to ensure that an airline like Pan Am could survive.[74]

Building the Chosen Instrument

Trippe and John Hambleton, a long-time Yale friend who was also an investor in Trippe's three aviation companies, went to Washington, D.C. soon after their Havana triumph to lobby for more lucrative overseas airmail contracts. Although in 1927 there was only $150,000 appropriated for foreign airmail, the winds shifted in Pan Am's favor in the aftermath of the findings made by President Coolidge's interdepartmental committee to encourage the development of commercial aviation in Central and South America. In November 1927, the members of the committee proposed creating air routes to the Southern Hemisphere that resembled those described earlier in Trippe's October speech to Pan Am's employees. In addition, they advocated more flexible rates of compensation that would be awarded not simply to the lowest bidder, but rather to the airline that the postmaster believed would be in the best position to serve U.S. interests—probably two or maybe just one airline would be the most efficient solution.[75] That the committee's findings strongly parallel Trippe's ideas is probably no coincidence. Hambleton was well connected in government circles and the committee included the Assistant Secretary of State Francis White, Yale class of 1913.[76] Just a few years later, the *New Republic* marveled at Trippe who had "so suavely and smilingly smoothed Pan-Am's path through the government's maze."[77] Who could have guessed in 1927 that this was the same Trippe who had been outmaneuvered by a group of dogsledders just a couple of years earlier?

To push forward legislation that eventually became the Foreign Air Mail Act in March 1928, Trippe focused on persuading those people who had resisted

Juan T. Trippe (far left), Irving Glover and T.H. Vane are shown seated at the West Indian Aerial Exposition Inc. (standing: C.S. Whitney) as Glover opened bids for two air mail routes- one to Puerto Rico and the other to the Canal Zone. (Source: Bettmann/CORBIS).

Hambleton's overtures. One prickly character was Coolidge's Assistant Postmaster General Glover, whom Trippe had first met during his days at Colonial.[78] Glover and Postmaster New both favored government backing of a single U.S. airline with international routes, but Glover particularly objected to granting a monopoly *and* high postal rates: "If you can't do it for that [a dollar a mile for airmail]," barked Glover to Hambleton, "others can."[79] Trippe quickly made an appointment with Glover and tried to convince him that supporting Pan Am was an economic *as well as* a diplomatic boon for the government. If Pan Am's flights attracted attention from South America, he assured Glover, the government might even make a profit with fees from foreign mails delivered to the United States. "We're not asking for a subsidy," explained Trippe. "We're giving the government a chance to make money, don't you see."[80]

Although this lucrative scenario was a stretch at the time, Trippe's entreaty probably softened Glover, who became much more receptive to Trippe's demands after returning from a postal tour of Europe where nations such as France were pouring millions of francs into French airlines in South America. Around February 1928, Glover had changed his tune. Testifying in a congressional hearing on foreign airmail, Glover admitted that, at least for a while, America's international aviation had to be subsidized to exist on a level playing field with European competition.[81] The assistant postmaster eventually became a partisan of Trippe's bid to have Pan Am monopolize international aviation from the United States. On July 11, 1928, Glover secretly allowed Trippe to look at a competitor's bid for FAM 6, a Caribbean route that ended in Puerto Rico. Just three days later, Pan Am was awarded that airmail contract.[82]

Although Trippe was a master at bending government policy to his will, it should be noted that Trippe's most persistent detractor in the government was the U.S. State Department. The department's quarrel with Trippe concerned his desire to monopolize international airline services from the United States. Traditionally, the State Department was careful not to antagonize foreign governments by forcing them to accept the services of only one U.S. business. Trippe was resourceful enough to cultivate relationships with the minority of State Department officials who did agree that a Pan Am monopoly was the most efficient way to promote U.S. interests (and to thwart foreign incursions in the Americas) via the airways.[83]

Preceding the passage of the Foreign Air Mail Act, Trippe not only lobbied Glover but also visited the designer of the 1925 Air Mail Act, Representative Clyde Kelly, who was also the sponsor of the Foreign Air Mail bill. Trippe's efforts certainly paid off when the bill passed. Instead of Glover's previous limit of paying $1 per mile flown, the Foreign Air Mail Act of March 8 allowed for a maximum of $2—and included 10-year contracts. Two months later, Congress appropriated $1.75 million for foreign airmail services—more than 10 times the amount allotted just a year before.[84]

The election of Herbert Hoover as president of the United States in November 1928 only seemed to accelerate the government's commitment to cultivating links with South America. Preceding his inauguration in 1929, president-elect Hoover made headlines by taking a good-will trip to South America that totaled more than 18,000 miles. Taking a train from Washington to California where he boarded a steamship, Hoover practically circumnavigated the Americas visiting four Central American and six South American countries between November and January. The goal, according to the *New York Times*, was to improve diplomatic and economic ties. The European powers, the *Times* explained, had recovered from World War I, and unless the United States redoubled its efforts to strengthen economic interactions with foreign countries in its back yard, "recession in prosperity would ensue."[85]

Not to be outdone, even by the president of the United States, Trippe and his wife accompanied Lindbergh on a well-publicized 7,000-mile airplane tour of South America starting on September 20.[86] The tour took them to Dutch Guiana via Puerto Rico. This was Pan Am's inaugural flight of FAM 6 (Miami to Port of Spain). The airline boasted of its ability to fly this route in four days, beating the steamship competition by two weeks.[87] Assuring the public's interest, Lindbergh was accompanied by his new bride Anne Morrow, the beautiful and literary daughter of the U.S. Ambassador to Mexico, Dwight Morrow. Trippe's instinct for obtaining favorable publicity was confirmed by the *New York Times*, which featured several pictures of the voyage, including a group portrait of the four world travelers getting ready to begin their journey from Miami.[88]

At this time, Lindbergh had been officially in the employ of Pan American Airways for six months. After their meetings in New York in 1927, Trippe and

Lindbergh met again in February 1928 during the Pan American Aerial Conference where the United States "redefined the Monroe Doctrine to include the air" (as well as the sea) and "set the stage for a single U.S. airline, Pan American Airways, to dominate international aviation in the Western Hemisphere."[89] Lindbergh was as enthusiastic as anyone about how airplanes might transform the Caribbean. "This territory is waiting for airlines," Lindbergh declared to the *New York Times*.[90] Despite their shared enthusiasm, Lindbergh politely refused Trippe's request in Havana that Pan Am formalize its relationship with the aviator.[91] But just the next month, Lindbergh made his first public flight for Pan Am, flying mail and passengers between Brownsville and Mexico City.[92] The patient wooing of Lindbergh for two years had paid off. Now, wherever the aviator went, Pan Am's name would follow.

Although much of Pan Am's future success was built upon the U.S. government's inclination to promote U.S. aviation interests through a single airline, in the late 1920s Pan Am was hardly the sole domestic or foreign airline with hopes of creating a hemispheric empire. Trippe had to navigate through an intricate and delicate web of financial, logistical, and legal obstacles to realize his goal. Never letting go of his monomaniacal obsession to subdue the competition, Trippe was nonetheless impressively nimble and creative in the means he employed to increase Pan Am's reach around the Americas.

First of all, lots of money was required to get Pan Am off and running. Trippe was blessed with the good fortune of having a colleague with the financial prowess of Richard Hoyt. Hoyt was then considered to be "Wall Street's aviation oracle, a gatekeeper of its insiders' paradise."[93] When Pan Am was just getting started, Hoyt took advantage of the growing interest in aviation stock with a clever scheme. In June 1928, speculation about aviation stocks was still rampant and Hoyt took full advantage of that investor enthusiasm. Hoyt first created a new corporation by making Pan Am's holding company a plural noun: Aviation Corporation of the Americas. This new corporation essentially acquired all the assets Trippe and Hoyt already possessed (in Aviation Corporation of America *and* Hoyt's old Atlantic, Gulf & Caribbean Airways). In this new guise, the old assets would back the sale of a new offering of stock to the public (a maneuver known as "watering down" the stock). Hungry investors were not asking questions. Stockholders of Aviation Corporation of the Americas even agreed subsequently to buy 90,000 additional shares that brought in $1,350,000.[94]

Armed with these kinds of assets, Trippe invested the money to facilitate Pan Am's bids for the lucrative foreign airmail contracts. Unlike his days at Colonial, Trippe now had free reign over a company that possessed the money and the personnel to promote his vision. One particular investment that paid off was funding Pan Am's agents to obtain tentative landing agreements with particular foreign governments *before* the U.S. Post Office even advertised airmail routes that would require those governments' consent. Trippe used these agreements to show that Pan Am could fulfill the requirements of the contract quickly.[95]

In the late 1920s, foreign and many domestic airlines were equally drawn by the lucrative potential of flight across the Americas. Although Trippe was often merciless with competitors from the United States, foreign competition offered different challenges. The largest obstacle to doing away with foreign airlines was national pride. The people of Central and South America were often very sensitive about Yankee imperialism, and Trippe did not want to create any unnecessary obstacles to obtaining his foreign air routes. For instance, in Mexico, the government mandated that only Mexican airlines could fly over Mexican air space. And in the 1920s, there was only one airline in Mexico: Compañía Mexicana de Aviacion, which was, ironically, founded by Americans. It was a tiny airline used principally to allow oil companies to distribute their payrolls on the Gulf Coast while avoiding the danger of bandits. After some negotiations with Compañía Mexicana, Trippe simply bought the airline in January 1929 and made it a subcontractor for Pan American. Because of the airline's strategic importance to Pan Am's expansion to the south, Trippe agreed to pay much more for the airline ($150,000) than its worth on paper ($5,000).[96]

In the next few months, Trippe pulled out every stop to secure a great prize. Just before leaving office at the end of the Coolidge administration, Postmaster New wanted a final great achievement. On March 2, 1929, New chose to award Pan American a mail contract (FAM 9) whose route passed through Panama to Chile and over the Andes to Argentina.[97] But this clear path through South America was suddenly obstructed in April by a competitor. Ralph O'Neill, a confident businessman and engineer who had also flown in World War I, decided to start an airline that would provide service to South America's east coast. Bringing on a generous investor who committed $1.5 million, an airplane manufacturer who donated 6 large sea planes, and a lawyer who had the ear of President Hoover, O'Neill incorporated New York, Rio & Buenos Aires Line (NYRBA) with high hopes of success. It was too bad for O'Neill that he had not learned from a former airline president who had just grappled with Trippe for control of the Caribbean: Basil Rowe of West Indian Aerial Express. After losing a bid to win FAM 6 despite his company's ability to fulfill the mail contract, Rowe reflected: "While we had been developing an airline in the West Indies, our competitors [Pan Am] had been busy on the much more important job of developing a lobby in Washington."[98]

Besides underestimating Trippe's agile lobbying powers in Washington, D.C., O'Neill committed an error that created a fatal flaw in the future of his business. Unlike Trippe, O'Neill started spending massive amounts of money on infrastructure in South America (thanks to his generous investor) *before* he was awarded a contract from the post office—one of the few risks Trippe never took. Trippe could also still count on his friend Assistant Postmaster Glover, who remained in office during the transition from the Coolidge to Hoover administrations and was only too willing to see Pan Am's competitors as threats to U.S. interests.[99]

NYRBA had already been awarded mail contracts from countries such as Argentina, but at a much lower rate than Pan Am would ever offer: $10 versus $25 dollars a pound to fly mail to Miami. Although NYRBA's fees attracted foreign governments, the business was hemorrhaging money. The needed remedy would be a contract from the post office, which seemed likely to O'Neill because NYRBA was already functioning as an airline in South America. But Trippe's lobbying helped to delay the advertising of what would be known as FAM 10, a contract for transporting the mail to the east coast Brazil. This delay allowed time for Trippe to lobby Postmaster Walter Folger Brown. The postmaster was feeling pressure himself because of the deficit the post office was accruing with foreign mail contracts (earning $700 weekly but spending $13,000). With a competitor such as NYRBA, the post office would never be able to pressure foreign governments to pay rates comparable to the $2 a mile fees now paid for foreign airmail by the United States. Brown decided to order NYRBA (as he had many airlines in the domestic sphere) to merge with its competitor in May of 1930. Driving a hard bargain, Trippe bought NYRBA, which had invested $6 million in its efforts, for little more than $2 million worth of overvalued Aviation Corporation stock. Soon afterward, Trippe won FAM 10. South America had been secured.[100]

The same month that Pan Am began its first regular scheduled passenger service—to San Juan via Belize and Managua—the chief of Latin American Affairs at the State Department succinctly described the important role Pan Am played in the U.S. government's plans in Latin America. In a memo of January 1929, Stokely W. Morgan wrote: "We have been moving heaven and earth to help Pan American Airways. This company is in an exceptional position in that the Department is very seriously and vitally interested in the success of its undertaking."[101] Although this memo's enthusiasm for Pan Am is unrestrained, Morgan could not have foreseen the increasing importance of Pan Am for U.S. interests all around the world. While international tensions were steadily increasing toward another global war, the utility of an American airline pioneering air routes that could have great strategic importance in the Atlantic and Pacific grew tremendously. Seizing this opportunity, Trippe would soon focus his attention on helping to create innovations in airplane technology that could help him to realize the almost unbounded aviation needs of the U.S. government in the 1930s and 1940s.

CHAPTER 3

C. E. Woolman and Delta Air Lines

The survival and eventual growth of Delta Air Lines during the tumultuous early days of the airline industry in the 1920s and 1930s was due in large part to the character of its director, Collett Everman (C. E.) Woolman. Simultaneously visionary and practical, risk-taking and fiscally prudent, a university-educated cosmopolitan who naturally blended into small-town Southern culture, Woolman somehow balanced these contradictory attributes and used them to create a prominent national business.

Long before transforming Delta from its initial identity as a crop dusting company in the Deep South to becoming one of the nation's leading airlines, Woolman was passionate about flight. The son of a physics professor at the University of Illinois at Champaign-Urbana, Woolman displayed an early aptitude for complicated and creative projects. As a boy growing up in Illinois in the 1890s, he is reported to have constructed a giant kite designed for passenger flight. (The prototype never got off the ground.) Less than a year after the Wright Brothers unveiled their own aerial accomplishments to an awestruck world at a 1908 aviation exhibition in France, Woolman decided to follow their example by attending the very first international aviation meet in Reims in 1909. His trip to France from Illinois was paid for by a summer of less-than-glamorous labor: cleaning the stables of 800 calves.[1]

The scene in Reims was bound to inspire any enthusiast about engine-powered flight. The *New York Times* reported that "aviation week" (as the French called it) featured 28 planes of American and European design flown by men with soon-to-be household names such as the American aviator Glenn H. Curtiss and the pilot who had just flown over the English Channel in July, France's Louis Blériot. The day before the official opening of the festivities on August 21, fifteen thousand spectators saw "Aeroplanes flying in straight lines or making wide turns or wheeling abruptly, traveling slow and fast and low and high." Though these "aeroplanes" were spectacular for the time, they were still small, slow, and fragile machines—so fragile, the *Times* noted, that the "aviators are chuckling to-night over an offer received by Augustus Post, Secretary of the Aero Club of America, from a Frenchman weighing 250 pounds, who says he will give $100 to each and every aeroplanist who will take him along in his flights."[2]

Undoubtedly, Woolman's week in France must have been gratifying to the young man who had dreamed of flying. On his voyage home, Woolman continued to indulge his interest in airplanes, thanks to a serendipitous encounter with American aviator Claude Grahame-White who asked Woolman to help him overhaul a plane engine he was planning to use during an upcoming aviation meet in Boston. Despite these unique experiences that could have tempted him to devote his attention to the nascent field of aviation, Woolman must have sensed that airplanes, however exciting, offered little certainty to those interested in a career in 1910. Prudently finishing the educational path that he had begun before traveling to Reims, he returned to the University of Illinois where, at the age of 23, he graduated with a degree in agriculture.

A Foundation in Agriculture

The next few years Woolman took solid steps forward in his agricultural career, moving south to farm in Mississippi and soon afterward becoming a manager of a 7,000 acre farm in northern Louisiana's Red River Valley.[3] Just a little while later, he joined the agricultural extension department of Louisiana State University (LSU) in 1913. Thanks to the Smith-Lever Act of 1914, which, according to the Democratic Party, was designed to convey "to every farmer in every section of the country, through the medium of trained experts and by demonstration farms, the practical knowledge acquired by the Federal Agricultural Department in all things relating to agriculture, horticulture and animal life," LSU received increased support for agricultural education.[4] The headquarters for this work was located in Monroe in the northeastern part of the state (the future headquarters for Delta). Woolman's proficiency in this job landed him a promotion to district supervisor of the northern Louisiana district in 1916. Despite his ever-growing responsibilities, Woolman still found time to learn how to fly a biplane.[5]

In 1916, Woolman returned briefly to Illinois to marry Helen Fairfield who was a home economics teacher in Champaign.[6] Over the next decade, the young agricultural educator quietly deepened his connections to the Monroe community by teaching farmers about the latest techniques to improve their farms and their bottom lines. Naturally gregarious with a disarming sense of humor, the northern-born agricultural expert found a home away from home in a Southern culture that valued his conversation skills, his quiet confidence, and his egalitarian demeanor.[7]

In 1921, a revolutionary agricultural technique was developed locally that offered great promise to eradicate the boll weevil, a little worm that had destroyed cotton crops for much of the previous decade. In the nearby labs of the U.S. Department of Agriculture located just east of Monroe in Tallulah, the etymologist Dr. Bert T. Coad explored ways to make the insecticide calcium arsenate both efficient and commercially viable. An aviation enthusiast himself, Coad successfully petitioned the federal government for funds to experiment with airplanes as a delivery mechanism for the poison. In the years following World War I, a glut of military planes, as well as pilots and mechanics, flooded the airplane market that provided the foundation for inexpensive crop dusting. Unfortunately, these warplanes were never well suited for crop dusting; in addition, the overabundance of cheap war surplus airplanes made it difficult for fledgling plane manufacturers to introduce new models of aircraft that were better suited for commercial and agricultural uses.[8]

Huff Daland Crop Dusting

Just a few years later, however, a bit of luck helped to make northern Louisiana the national center for the nation's crop dusting activities. In 1923, Coad met George B. Post the vice president of Huff Daland Company, specializing in military aircraft and based in Ogdensburg, New York. While he was on his way to Texas to demonstrate the company's new biplane trainer, Post made a forced landing in Tallulah. After meeting with Dr. Coad, Post quickly became enthusiastic about the potential of crop dusting to jumpstart Huff Daland's moribund sales. Huff Daland then carefully entered this new industry by consulting with Coad and similar researchers in the federal government. When it became clear that no other companies were moving to develop airplanes for crop dusting, Coad and Woolman together persuaded Huff Daland to dedicate a whole new division to this endeavor. In 1924, Huff Daland Dusters, Inc., (with Post assuming the role of president and retired army pilot Lieutenant Harold R. Harris as operations manager) located its headquarters in Macon, Georgia, which offered two promising elements for the company's success: a variety of crops located around one of the few usable airfields in the region.[9]

When Huff Daland first began its operations in Georgia, C. E. Woolman had already spent 12 of his 34 years working intimately with the farming, educational, and business communities in and around Monroe, Louisiana. In 1925, Woolman left his secure job to join Huff Daland Dusters. As the company's first operations manager Lt. Harris recalled, Dr. Coad recruited Woolman to become a vice president of Huff Daland Dusters because the company had been "unable to sell its services with the personnel it had." Woolman used his agricultural expertise to supervise the company's work with pesticides. More importantly, he employed his conversational skills and local connections to market Duster's services. "Coad was right," Harris concluded, "C.E.W. was a great salesman."[10]

But Woolman was much more than an ingratiating salesman: he was a gifted entrepreneur who could expand small markets as well as develop new ones. The first step that Coad and he took together was to move Daland's headquarters from Macon to Monroe. Monroe had many obvious advantages for the two Daland executives whose combined experience working in northern Louisiana totaled 29 years. But relocating to Monroe was much more than a matter of convenience and cultivating familiar terrain: Monroe was situated in the middle of boll weevil territory, and farmers were likely to respond very favorably to Daland's solution to their insect problem. The only thing the city lacked was an airfield. Luckily, Woolman was not the only prominent local resident who was enthusiastic about aviation. With the backing of some prominent business leaders who shared Woolman's interest in the future of aviation, the cities of Monroe and West Monroe committed themselves to building a public airfield if Huff Daland agreed to stay in Monroe for a minimum of three years. The result was Monroe's Smoot Field.[11]

Woolman and his Huff Daland associates now possessed the necessary ingredients for success in the short-term: a unique product, a large potential market, great public relations, and the infrastructure to support their operations. During the next couple of years, the company grew beyond its initial inventory of 18 planes to better serve an ever-widening radius of farm country extending from Arkansas to Mississippi and Georgia and eventually moving as far west as California.[12] In these two years, Huff Daland Dusters dominated the cotton crop dusting market. They were so successful that during the late 1920s, the company housed the "largest privately owned aircraft fleet in the world."[13]

Although business was going well in the American South, Woolman's global vision and experience inspired an ingenious marketing innovation to expand the calendar year for crop dusting by aiming Huff Daland Dusters services farther afield. The first step south took Huff Daland to Mexico in 1925. The next step, however, was not as close or obvious. In 1926, Woolman headed a

Early Huff Daland Duster with Liberty engine applying calcium arsenate on cotton in 1924. Flown by Lt. Harold R. Harris. (Source: Delta Air Lines).

marketing expedition of sorts to investigate business possibilities among "the cotton growing estates of the Peruvian coastal valleys" where he successfully sold the idea of crop dusting to Peruvian farmers.[14] Peru's cotton crops were a promising new area for Huff Daland's operations because their growing season preceded that of the Mississippi Delta region by several months. Unable to cover such a distance directly, five of the Huff Daland airplanes were disassembled, crated, and shipped to Peru. Within a year after operations began in two of Peru's "most fertile valleys" in 1927, Huff Daland Dusters worked over "seven of Peru's most heavily farmed areas."[15] Woolman employed his sense of humor and cosmopolitan sensitivity to the local culture by approving the use of a Spanish nickname for Huff Daland's Peruvian operations: *La Llama Voladora* ("the flying llama").[16] Each of the crop dusters sported a logo that depicted a man seated on a llama, thus implying that the dusters were as reliable and useful as the famous pack animal.

Here we see a common pattern in the early evolution of an industry. The structure of demand is uncertain, and early entrepreneurs place bets on different customer segments that they believe will provide the growing demand that will enable them to build a viable business. As we saw in the previous chapter, Trippe first focused on transporting wealthy individuals to their vacation homes. When this demand proved limited and competition for it intense, he turned his attention to serving companies that might need documents transported across difficult terrains on a timely basis, then on delivering mail for the Post Office Department, both domestically and internationally, and finally on international passenger travel. Trippe's initial choices may have reflected the customer segment he knew best—wealthy families that his friends at Yale belonged to. Similarly, Woolman's early choices reflected the segment he knew best—farmers whose crops were threatened by disease for which crop dusting provided a solution. As we see in the airline industry, variations in the backgrounds of individual entrepreneurs can have important consequences for the evolution of consumer demand, and the business models that the individual leaders craft.

While Woolman was making inroads into Peru, Huff Daland's parent company Keystone Aircraft was undergoing major changes that dramatically altered the dusting company's future. During this time of rapid consolidation in the airline industry, Keystone was acquired by a group of Wall Street financiers, including Richard F. Hoyt, who was deeply involved in the early years of Pan American Airways. Huff Daland's pioneering efforts in Peru caught the attention of Hoyt and other executives, including Juan Trippe. With his eyes focused on laying out the groundwork for creating a South American network of air routes, Trippe wanted to use Huff Daland as the means to secure a foothold in Peru for Pan American. So, in September 1928, Woolman, now as an employee of Pan American (through Hoyt's acquisition of Huff Daland), participated in negotiations with the Peruvian government, which resulted in Peru granting a charter to a subsidiary of Pan Am, Peruvian Airways, Incorporated.[17] Although they were certainly grateful for Woolman's effective leadership in a foreign land, Pan Am's executives maintained no romantic attachment to Woolman or Huff Daland for their pioneering efforts. So, almost as soon as Peruvian

Airways had been created, Hoyt decided to sell Huff Daland and told close associates of Woolman that he wanted to "close out the dusters to the pilots on any terms (or to anybody else, as far as that goes)."[18]

Although this was an abrupt move, it did not catch Woolman unprepared. Apparently sensitive to the volatile nature of the early airline industry, Woolman had already created a contingency plan to acquire Huff Daland if it were put up for sale. This plan would be implemented with the help of two associates: Vice President Harold R. Harris and comptroller Irwin Auerbach. When Harris heard of Hoyt's desire to "sell out for forty thousand dollars, and all notes if necessary," he was eager to share the good news with Auerbach: "It looks like a golden opportunity for us to do some good for ourselves."[19]

Thinking that the trio would push forward together to buy Huff Daland's American and Peruvian assets (the latter was valued at approximately $15,000), Woolman returned back from Peru on October 24, 1928 to find that all was not going well: Auerbach was trying to acquire Daland by himself. In official correspondence sent to investors before Woolman's return, Auerbach represented Woolman simply as a potentially interested buyer. After confronting Auerbach, whose answers were often incomplete or evasive, Woolman wrote to executives at Keystone Aircraft to help him unravel this intrigue. Luckily for Woolman, the comptroller's maneuvers had aroused the suspicion of many of Huff Daland's investors and employees. When the Keystone executive confirmed the betrayal, the Monroe community refused to deal with Auerbach. Although Auerbach's scam could have turned investors sour on the future of the crop dusting company, Woolman was able to raise the necessary funds from local bankers and planters as well as some of Huff Daland's major executives. Once again, Woolman's trusted role in the Monroe community—along with his celebrated successes in South America—provided help to his aviation business during a difficult time.[20]

This episode in Woolman's career demonstrated the interesting role that merger and acquisitions can play in shaping and reshaping the opportunity structure for early entrepreneurs in an industry. Being successful in part requires quick maneuvering to remain in a position where you have some control over your own destiny—having a network of trusted friends and acquaintances, who will finance you in a pinch, is one of the assets that enables entrepreneurs to successfully navigate this merger and acquisition induced turbulence. Trippe was also a master of building and nurturing influential and powerful relationships that enabled him to create three airlines in a relatively short period of time. Access to capital is crucial in any industry's start-up phase, and capital is often initially secured through angel investors who are, more often than not, personal friends or acquaintances of the entrepreneur.

Delta Air Service

On November 18, 1928, Huff Daland Dusters became "Delta Air Service, Inc." with Woolman as first vice president. Though he held the title of vice president, Woolman served as the de facto head of the company. Two major new investors joined Woolman in directing the company: D. Y. Smith, a local planter, became Delta's president; Travis Oliver, an important Monroe banker who had worked with Woolman to find investors in addition to making a large personal investment himself, became treasurer. Harris remained with the company as second vice president. Catherine Fitzgerald, a secretary who had moved from Keystone's headquarters to Monroe in 1926, suggested the new name of the company in honor of the Mississippi Delta region where Huff Daland had done so much of its business. Although Delta would immediately resume the agricultural work that had made Huff Daland Dusters so prosperous, Woolman had plans to enter into new business territory.[21]

As Delta Air Lines historians W. David Lewis and Wesley Phillips Newton point out, it would be difficult for this fledging company to "survive in an industry that was already experiencing formidable pressures towards consolidation."[22] Those pressures, in fact, may have been the motivating factor for Woolman to move so quickly and decisively to claim some territory in the airline industry. And although he could not have predicted the turn that government regulation would have taken in the near future, Woolman's eagerness to move from crop dusting to passenger service gave Delta an important legal foothold in the airline industry that proved to be crucial to its survival.

C. E. Woolman's Peruvian adventure in 1928 provided inspiration for Delta Air Service's entrance into passenger service. Indeed, if airplanes could function well in underdeveloped nations with air routes disrupted by the vertical walls of the Andes (airplanes of that era could not yet get over most mountains), they certainly could fly over the more economically vibrant and geographically flat areas radiating from Monroe, Louisiana. Woolman's good relations with Monroe's political and business leaders ensured that local conditions were ripe for Delta's take-off plans. But to be successful, Woolman also needed to understand and anticipate the important regional and national influences around 1928.[23]

Early Passenger Service

Three years after the passage of the Contract Air Mail Act of 1925 (the Kelly Act), the Post Office Department had awarded airmail contracts for "feeder lines" that connected smaller cities (generally arranged in a north-south direction) to the already-established transcontinental route (generally progressing in an almost straight line from New York to San Francisco via Chicago).[24] A quick look at a map of the early contract mail carriers at the time shows that the South was largely left out of the loop of this feeder system. Only one major airline (Florida Airways) provided airmail service, and this connected Atlanta to Miami. Woolman saw an opportunity for an east-to-west airmail route in the deep South.

The regulatory discretion of those who ran the post office encouraged the development of passenger service. Specifically, President Calvin Coolidge's Postmaster General Harry S. New openly favored large companies that had the financial means to acquire the largest aircraft available. Eventually, New hoped, companies with this increased capacity would derive larger amounts of revenue from passenger service. The young airline industry responded according to New's incentives—with small private companies being challenged by larger conglomerates. The first major airline company to emerge was Transcontinental Air Transport (TAT) (forerunner to TWA), which was formed in 1928 and began operations after a year of careful planning. Before the age when airlines could fly at night, transcontinental carriers (United and American, companies that were built from the merging of many small airlines, soon followed TAT) organized a combination of air and rail service.[25] TAT relied exclusively on passenger service, a strategy that, at least in the late 1920s, proved to be wholly unprofitable. After 18 months of passenger service, TAT lost $2,750,000.[26]

Despite some spectacular failures in the beginning of the airline industry, passengers and capital began to flock to the airlines. One source of the growing public interest in airlines can be attributed to the well-publicized aeronautic feat of Charles Lindbergh. Although some academics point to the fact that investment in airlines was growing before 1927, Lindbergh's accomplishment certainly did not hurt the cause of the airline industry: between 1927 and 1929, investment in airline stocks tripled. The public's skyrocketing enthusiasm for flying may have further encouraged investors to place their faith in the new industry. From 1926 to 1929, the number of revenue miles flown grew from 4.3 million to 22.7 million, and the number of airline operators increased from 13 to 38.[27] This occurred despite the many discomforts passengers had to face in uninsulated, unpressurized, and noisy planes that flew too low to avoid the often nightmarish turbulence at higher altitudes. The Lindbergh fever of the late 1920s was not just a distant national story for the people of Monroe. In October 1928, Lindbergh visited the nearby city of New Orleans, which further invigorated the region's enthusiasm for aviation.[28]

Given the generally well-known difficulty of turning a profit from passenger service alone, it is somewhat surprising that C. E. Woolman decided to begin Delta Air Service as a passenger-only airline. Woolman had probably studied the idea of making Delta an airmail carrier as well, because a federal airmail survey completed in June 1928 assessing the viability of a route through Monroe was found posthumously among his papers. (The entire route included Shreveport, Monroe, Jackson, Meridian, Tuscaloosa, and Birmingham.) A few months after the survey was conducted, Woolman also asked for advice from a Minnesota airline that did concentrate on passenger traffic.[29] Perhaps Delta's lack of cash made it difficult to start an airmail service that would meet the expectations of Postmaster New.[30] In addition, a local airline, St. Tammany-Gulf Coast Airway (established in 1927), already offered airmail service between Atlanta and New Orleans. This route was expanded when St. Tammany became part of Gulf Air Lines, which added a New Orleans–Dallas connection in May 1928.[31] Woolman's decision to focus on passenger traffic could have also been encouraged by mainstream press such as the *New York Times* that acknowledged the difficulty of making money from passenger traffic in the past but also predicted a very rosy immediate future.[32]

Once the decision was made to orient the company toward passenger service, Woolman moved quickly. In early 1929, Woolman was fortunate to meet a young and wealthy businessman from a successful Kansas City family by the name of John Fox. Fox lived in the adjacent town of Bastrop, Louisiana and had learned how to fly in 1926. After a half-hearted attempt at beginning his own airline, he sold the few assets he had acquired to Woolman and became the principal stockholder of Delta Air Service, having acquired $55,000 worth of the company's shares. The deal helped Delta acquire two of Fox's airplanes and another second-hand plane he had on order without increasing its debt substantially.[33] The model of Delta's first two passenger airplanes was the "high-wing, single-engine, six-passenger, enclosed-cabin Travel Air [6000-B]" built by the Travel Air Company of Wichita, Kansas. The plane, described by Monroe's press as "the last word in airplane construction," boasted a range of 500 miles and a top speed of 130 mph.[34] Woolman's good fortune to have met with Fox was further enhanced by the fact that the Travel Air Company—whose co-founders Walter H. Beech and Clyde V. Cessna would become famous pioneers of airplane manufacturing—made some of the best planes of the late 1920s.[35]

Delta Air Service advertisement for passenger service, 1930. (Source: Delta Air Lines).

Delta Air Service's first passenger flight took place on June 17, 1929 from Dallas, Texas to Jackson, Mississippi, with stops at Shreveport and Monroe, Louisiana. Extensions to this route were made soon afterward: further to the east at Birmingham and to the west toward Fort Worth. Nine months later, Delta Air Service finally added Atlanta to its list of destinations on June 12, 1930.[36] The company offered flights 2 times a day between Atlanta and Fort Worth, a route that could be completed in a total of 10 hours and 15 minutes.[37]

The route made sense for many reasons. Dallas was a large city in the South that did not have any direct east-west passenger routes that connected it to Atlanta and, by extension, the eastern seaboard. Dallas was also connected to a national network of airlines through the mail and passenger service provided by National Air Service that flew between Dallas and Chicago.[38] This original route also went through much of the territory that Woolman had already become familiar with through the crop dusting work of Huff Daland.

The early financial returns on Delta's passenger service were disappointing. While Delta made a $20,000 profit for its dusting operations between November 1928 and December 1929, Delta's passenger service revenues were $32,000 in the red. Despite this bad news, Woolman continued with his plans to expand operations to Atlanta in the hopes of garnering more passengers. To compensate for some of these poor returns, Woolman decided to petition the post office for a mail route. He wrote optimistically to his head of operations Harold Harris in April of 1930: "with the Watres Bill now before Congress we have great hope of coming in . . . for air mail over the run."[39] The McNary-Watres Bill (approved on April 29, 1930) was designed to encourage further growth of passenger airlines by paying airlines for the overall size of their airplanes rather than the amount of mail they carried. Airlines could almost double their money by selling tickets for passengers who would fill any empty space not taken up by the mail.[40]

Navigating New Government Policies

Although this bill in principle could have been a boon for Delta, in practice it almost became the first step in Delta's demise. The bill gave "dictatorial" powers to the postmaster over the future of the airlines. Postmaster Walter Folger Brown's first preoccupation with the future of passenger and airmail service was to foster the development of larger airlines with deeper pockets and longer routes that could offer some stability to the struggling industry. He did not want to eliminate competition altogether, but he wanted to foster an airline passenger system that the public could depend on for years to come. In 1930 when Brown became postmaster, there were 44 small airlines in existence of which Delta was probably among the smallest.[41] Brown was most concerned about creating a transcontinental air service, and he believed that a coast-to-coast route could not be run by multiple companies.[42]

Southern businessmen who supported Delta's goal of taking on mail for the federal government petitioned the post office on Delta's behalf in May 1930, asking that the fledgling airline receive an airmail contract.[43] Praising the company for its impeccable safety record and its sound leadership, they had no idea that Brown was moving quickly to change fundamentally the way the airline industry was organized, a move that threatened to put small companies such as Delta Air Service out of business.

At the same time Delta's petition reached Washington, the postmaster called a meeting of most of the nation's airlines to discuss how to administer the new airmail contracts under the McNary-Watres Act. The act cancelled all existing air contracts and placed significant stipulations on the rebid process. The stipulations were mostly intended to weed out the smaller carriers. Airlines were required to have daily route schedules covering at least 250 miles and have 6 months of demonstrated night flying. The mail subsidy was also cut almost in half. A few years later, this meeting came to be known as "the spoils conference" in which certain airlines were favored (receiving the spoils of the airmail contract awards) and others were forced to merge and/or sell out.[44] Woolman did not even receive an invitation to the "conference," perhaps because Delta, like many other small independent airlines, was too small of an operation to receive much serious attention from Brown. Luckily, Woolman received a telephone call alerting him to the meeting, and he immediately went to Washington to defend Delta's interests.[45]

If not frantic, Woolman certainly became frenetic and used his creativity to find a way out of this trap. Although Brown had conceded that Delta had a legitimate claim to have "pioneered" passenger service between Atlanta and Dallas, Woolman could see very clearly that Delta was not going to be permitted to survive in its present state—it would have to become affiliated with a larger airline company. As an entrepreneur, Woolman had to become flexible in his vision for the company, and he quickly pursed a partnership arrangement that would allow Delta to survive. Woolman first approached Eastern Air Transport, which was expecting to receive rights to fly between Atlanta and New Orleans and explored the possibility of subletting that route to Delta. When Brown became annoyed with Eastern's directors, he decided not to award the Atlanta-New Orleans route to them; instead, he gave the southern route to Aviation Corporation (AVCO), an aviation holding company (affiliated with American Airlines, not Pan American).[46]

A company that could be described as Delta's opposite, AVCO began with working capital of $35,000,000 a few months after it was formed in 1929 and set out to buy airlines from all over the country.[47] With these funds, AVCO embarked on a major acquisition spree, acquiring, in less than a year, almost 80 different aviation businesses, including 5 different airlines and 3 holding companies.[48] One of these airline holding companies was Southern Air Transport, which itself had bought, before being acquired by AVCO, St. Tammany-Gulf Coast Airway, Delta's neighbor based in New Orleans. Because St. Tammany's service had preceded Delta's, AVCO was awarded St. Tammany's "pioneer" service between Birmingham and Atlanta, to which Postmaster Brown added Birmingham to Fort Worth under an arbitrary "extension principle."[49]

Woolman traveled to New York to negotiate some sort of deal with AVCO. He met with the current and former presidents of AVCO and received a verbal agreement from them to buy a controlling interest in Delta and then to permit Delta to fly between Atlanta and Fort Worth as a subsidiary of AVCO. Unfortunately, internal wrangling between the postmaster and the Commerce Department invalidated the "extension principle" that had allowed Brown to award AVCO the route between Birmingham and Fort Worth. According to Delta historians Lewis and Newton, "the clause in the McNary-Watres Act granting to the postmaster general authority to extend existing routes and award them to favored companies could not be construed as empowering him to grant to an airline an extension longer than the route that it had flown under previous airmail legislation."[50] AVCO now had to competitively bid for the route, which left Woolman and his fledging airline with no more immediate options.[51]

With Postmaster Brown and the representatives from the nation's larger airlines cutting deals with little regard to the well being of smaller competitors, Woolman and other Delta executives had come to the conclusion that the best thing they could do was to sell to AVCO. Using what little influence he had through his connections to Louisiana Senator Edwin Broussard, Woolman tried to pressure the postmaster to ensure that Delta would be compensated fairly for the route that it had flown successfully. Maintaining service until the very end—perhaps hoping to discover an unseen foothold in the airline industry—Woolman's airline stopped all flights only when AVCO won the competitive bid and started flying on October 1, 1930. AVCO did finally pay Delta around $105,000, which, according to Woolman, was approximately half the amount Delta had invested in its Atlanta to Fort Worth route.[52] That marked the end of Delta Air Service's short history of passenger service.

In the end, the McNary-Watres Act eliminated many small, regional carriers from the bidding process and set the stage for Brown to award the bulk of mail contracts to three primary carriers who had the capability to traverse the continental United States. United won the northern routes, TAT won the central routes, and AVCO's American Airways won the southern routes.[53] Here we see the defining influence of government regulations, and the specific regulatory preferences of Postmaster Brown, on the evolution of this industry. The impact of government intervention has been felt by many industries, most notably in telecommunications, transportation, healthcare, and banking. For example, Christopher Marquis discusses how the shape of the U.S. banking industry was significantly influenced by regulations that restricted interstate branch networks in favor of local banks.[54]

Starting Over—A New Delta

Perhaps determined not to give up on a business in which he had displayed such acumen and for which he had much affection, Woolman regrouped to form a new company. He went back to crop dusting and bought planes and equipment from AVCO's Southern Air Fast Express for $12,500. This sale included an agreement that Woolman could continue to use the name "Delta Air Service" so long as the company refrained from transporting passengers. Woolman then acquired a new charter from the state of Louisiana (under the name of Delta

Air Corporation) that allowed it to pursue a wide array of activities related to aviation—from aircraft manufacture and maintenance to flight instruction. Unfortunately, along with the spoils conference came the Great Depression, which forced Woolman and his associates to come up with creative ways to make money in the airplane business and to become as frugal as possible.[55]

Farmers were the primary clients for crop dusters, and farms were doing very poorly when Delta Air Corporation began on December 31, 1930. Several months earlier, Woolman sent this discouraging assessment to his colleague Harris who was working with the dusting operations in Peru: "with low cotton prices the farmers are completely *broke*."[56] Luckily for the farmers, but not for Delta, the winter of 1930–1931 brought unusually cold temperatures down South, thus almost wiping out the problem of the boll weevil. In addition, this market now had regional competition—Curtiss Flying Service of Houston—that was charging farmers a meager 35 cents per acre dusted. Woolman persevered nonetheless, repeating his Latin American strategy by winning a dusting contract in Mexico.[57] Woolman and Harris decided to liquidate Huff Daland's Peruvian assets, which were sold in 1931.[58] In the same year, Dr. Coad came over to Delta to run the company's dusting operations (a position he kept for 35 years). But no matter how good Delta's dusting personnel were, agricultural business in the 1930s was not going to keep Delta Air Corporation solvent.

Staying within the category of "services associated with airplanes," Woolman pursued every possibility conceivable to generate badly needed revenues. He was hopeful about Delta having acquired a franchise from the Curtiss-Wright company to sell the company's goods in Louisiana and Mississippi. Having also become the company that managed Monroe's Selman Field (which replaced the original Smoot Field), Delta could generate more money through servicing aircraft (services which included repairs, routine maintenance, renting hangars, and conducting mandatory inspections). Delta also offered flying lessons for locals who occasionally ventured into aviation and helped private and government agencies conduct intermittent photographic surveys of the areas around northern Louisiana. Swallowing his pride, Woolman even arranged for Delta to work on the planes of its archrival, AVCO's American Airways.[59] With the profit margins slim at best for any of these activities, Woolman was sustained during this time in the aviation wilderness with the generous financial assistance of local banks whose services were often secured with the help of Delta's treasurer Travis Oliver.[60]

Despite the bleakness of the times, aviation enthusiasts continued to push the boundaries of what was possible. In 1931, Wiley Post and Harold Getty flew around the world in 8 days (covering a northern route totaling 15,500 miles) and were followed in 1932 by Amelia Earhart who became the first woman to fly solo across the Atlantic.

Changing Government, Changing Fortunes

A transition in the U.S. government brought about a welcome change in Delta's fortunes. In the first third of the 1930s, the government policy toward business was shaped by President Herbert Hoover's vision of an "associative state," which would, in theory, defend "American individualism" against the extremes of an economy dominated by large monopolies as well as against the constraints of an interventionist government. Instead, business would work together to foster "cooperative institutions designed to provide a voluntary network of self-regulation." The federal government would "prod them to put aside selfish impulses," but its authority to force any specific outcome would be limited so as to encourage "private solutions to [private business] problems."[61]

The succeeding administration led by President Franklin Roosevelt saw the actions that had been taken by Postmaster Brown to shape the airline industry as extremely coercive—and representative of the abuses of power in government and business that helped to perpetuate the agony of the Great Depression. In January 1934, Senator Hugo Black of Alabama led a committee to investigate Brown's reshaping of the air transportation industry. Opponents of the spoils conference of 1930, such as C. E. Woolman, were asked to testify in Washington.[62] By February, Black announced the conclusions he had drawn from the testimony in a radio address, which was transcribed in the *New York Times*: "It was never intended by patriotic citizens that this governmental aid [Black cited a figure of $58,000,000 in subsidies] should be diverted by collusive agreements into the pockets of favored bankers, brokers or stock manipulators, politicians and lobbyists."[63] The next day, Roosevelt cancelled all existing mail contracts and ordered the army to take over those routes in a few days.[64]

This ambitious move by the government turned out to be a fiasco for the Roosevelt administration, but it became an opportunity for Woolman and other small independent airline operators like him. The army was not prepared to take over airmail services in a matter of days that had taken years to build. A number of planes were damaged and approximately a dozen pilots killed in accidents that took place during the next few weeks. In addition, airmail costs rose from 54 cents a mile (under Postmaster Brown's regime) to $2.21 a mile over an airmail system that had shrunk from 27,000 to 9,000 miles of routes.[65] After receiving substantial criticism from the press and Republicans in Congress, Roosevelt cancelled government control of airmail.[66] There would not be, however, any return to the status quo *ante* Postmaster Brown. With the Air Mail Act of 1934 (co-written by Senator Black and passed on June 12), important pioneers of the aviation industry who had participated in the now infamous spoils conference were permanently banned from aviation. The act also forced all of the large holding companies that were involved in both airlines and airplane manufacturing to break up. Again, we see the extraordinary impact regulation can have on the structure of a nascent industry. Just as the 1930 intervention spurred consolidation and concentration in the industry, the 1934 regulation spawned the growth of smaller airlines. Although big airlines such as United and American remained more or less intact with largely cosmetic changes, there was room for more competition from smaller companies, which contributed to bringing down the price of airmail contracts (along with an administration that was less willing to subsidize airline costs). Winning bids for airmail contracts under the Air Mail Act were, on average, 40 percent lower than between 1930 and 1934.[67]

Technological Innovation in Airplane Design Drives Progress in Airline Services

Although the prospects for a prosperous future in the airline industry may have looked grim at this point (and immediately following the changes of June 1934 profits in the industry as a whole did go down), technological advances in airplane manufacturing alleviated this situation.[68] Some of the biggest advances in the 1930s were spurred by a well-publicized tragedy. In 1931, the famous football coach Knute Rockne died in a plane crash, which spurred the government to ground the wooden plane he was flying in (the Fokker trimotor) because of structural failure.[69] In response, the American airline industry spurred the manufacture of sturdier metal planes. The immediate result was the 10 passenger Boeing 247 (1933) and the 14 passenger DC-2 (1934) made by Boeing's competitor, the Douglas Aircraft Company.[70] Despite the fact that the DC-2 lagged one year behind Boeing's metal plane, the Douglas aircraft was much more popular with consumers. That popularity stimulated DC-2 production to a remarkable rate for the time: 10 planes per month.[71]

Although Woolman certainly would have liked to restart Delta's renewed passenger service with the best metal planes, he would have to make due with what he could afford at the time. In entrepreneurial fashion, Woolman looked at the situation as an opportunity, rather than an obstacle and began to move forward with his reconceived airline. After having been awarded the Charleston, South Carolina to Dallas airmail route by outbidding American Airlines 24.8 to 43.5 cents per pound of airmail, the rechristened Delta Air Lines began flying passengers on July 4, 1934. Delta also renamed that air corridor "The Trans-Southern Route," a moniker that seemed as tailor-made for Delta's consumers as had been "*La Llama Voladora*" in Peru. The plane used for this route was the seven-passenger, fabric covered, high-winged Stinson-Ts. These were planes that had been used by American Airlines—machines that the company was eager to be rid of as it began to buy larger aircraft. Delta bought them for $5,000 each: approximately a fifth of their original value. Within 5 months, Delta had transported 1,464 people on its southern route.[72]

Woolman's strength as a business man and an entrepreneur was his ability to recognize opportunities in their infancy stage and take them to the next level. While it is often easier to allow the stringent regulatory nature of an emerging industry, such as the airlines, to dictate the future of one's business, Woolman pursued the more difficult path. Instead of bending to regulatory pressures, he attempted to bend them in his direction. When this did not work, he adopted a new approach. As government policies and competitive forces changed, his nimbleness and adaptability enabled him to quickly reconceive his business plans. Reinventing Delta time and time again was Woolman's brilliance as an entrepreneur.

During the next couple of years, Delta barely broke even. But other measures during this time were far more promising. Gross revenues almost doubled from $245,000 to $431,000 with passenger revenue miles paralleling this growth, moving from 1.2 million to 2.3 million between 1935 and 1936.[73] This growth must have been very encouraging to Delta's new principle stockholder and president, Clarence E. Faulk, a local entrepreneur who filled in the void left by John Fox after his departure from Delta in 1932. Woolman convinced Faulk to grant a personal loan of $150,000 to finance Delta's acquisition of 3 Lockheed 10-B Electras, the first of which were delivered in December 1936. The Lockheeds' top cruising speed of 190 mph was quite an improvement from the 100 mph of the previously used Stinson Ts.[74] More importantly, as airline historian R. E. G. Davies points out, the Lockheed, which seated 10 people, may have been the best fit for the times because the passenger traffic levels of the mid-1930s rarely filled the seats of the 14-passenger DC-2 on a regular basis.[75]

New Regulations Drive Safety

In 1938, the federal government changed the airline business landscape yet again when the Roosevelt administration won passage of the Civil Aeronautics Act that established a five-member supervisory board called the Civil Aeronautics Authority (CAA). (The CAA would be replaced by a stronger Civil Aeronautics Board [CAB] in 1940.) The act enjoyed considerable support in Congress because it promised to establish some order to an industry that had been rocked by tremendous regulatory and economic changes during the previous decade. Harkening back to some of the informal elements of Postmaster Brown's implementation of Hoover's "associative state," the CAA wielded more formal centralized power over the airline industry than any agency had before. Among its many regulations, one of the most sweeping was a provision that no airline could perform airmail or passenger service without having received a "Certificate of Public Convenience and Necessity (CPCN)" from the CAA. After they acquired these certificates, airlines were even prohibited from abandoning routes without explicit permission from the government.[76] Quick to respond to such regulatory changes, Woolman made sure that Delta became the first airline to receive a CPCN (even though the government had not precisely defined what exactly constituted "public convenience and necessity.")[77]

Reacting perhaps to the hard lessons learned in the aftermath of the 1930 spoils conference, Woolman seemed to have anticipated that the CAA would bring with it a heightened scrutiny of the safety record of the airlines. Woolman directed Delta's chief pilot to invigorate the company's program of pilot training. The strategy paid off. While other airlines suffered a number of accidents in the late 1930s and early 1940s—including a few widely publicized crashes that horrified the reading public—Delta survived this era with an exemplary safety record.[78] This safety record contributed to the company's continuing success through the end of the 1930s and early 1940s. There was so much demand for Delta flights that Woolman decided to buy the larger DC-2s and

C. E. Woolman, Delta's principal founder and first CEO, 1940s. (Source: Delta Air Lines).

DC-3s (with 14 and 28 passenger capacity, respectively) in 1939. Although Woolman's decision to use DC-3s definitely aided its growth, it was hardly unique in the airline industry: "by 1939, 90% *of the world's* airline traffic was being carried by these aircraft."[79] The universal success of the DC-3 (due in large part to the efforts of C. R. Smith of American Airlines who will be discussed in chapter 5) presaged the post–World War II era in which technological advances in aircraft began to decrease in significance as a distinguishing competitive factor between the airlines. Indeed, Woolman knew that a shiny bunch of Douglas aircraft would not, by themselves, ensure a strong future.

Moving to Atlanta

As innocuous as they might seem, choices of where to locate the company headquarters had a significant influence on the fortunes of the early entrepreneurs in the airline industry. Woolman's final major decision on behalf of Delta before World War II was to move the airline's headquarters to Atlanta, which was approved by the company's board of directors on March 1, 1941. Although this proved to be an extremely important and helpful change that provided a solid foundation for Delta's growth for the rest of the twentieth century, Woolman was uncharacteristically hesitant to commit the company to what may have seemed like an inevitable course of action. By 1940, parts of Delta's operations had already migrated eastwards away from Monroe. Many of the company's pilots and co-pilots had established permanent residences in Atlanta and most of the company's maintenance operations were also based there. As Delta historians Lewis and Newton explain, moving from Louisiana to Georgia "was a logical arrangement because Atlanta was the hinge of the system, where the eastern and western divisions met."[80] On the other hand, Woolman had spent almost all his adult life in and around Louisiana, and Monroe in particular had been an especially friendly and familiar place for Delta to do business.

But by 1940, Monroe may have done all it could do for Delta.[81] The financial and intellectual resources available at Monroe were becoming less and less useful for a company with an increasingly national profile. Although Monroe took in steady revenues as a major player in the cotton and natural gas markets, it seemed to have reached an economic plateau that would have difficulty following Delta's slow but steady growth. For example, when in the summer of 1940 Woolman wanted to finance the purchase of 6 DC-3s at the cost of $115,000 each, the expenses were covered by increasing the amount of Delta's stock and by a $500,000 loan from a creditor in Atlanta.[82] Careful scrutiny of Delta's finances by Atlanta underwriter Richard Courts revealed that, unfortunately, previous stock transactions undertaken by Delta in 1939 had been in violation of rules set out by the Securities and Exchange Commission (SEC). The SEC was a relatively new agency founded in 1934 whose rules were unfamiliar to most people in small towns like Monroe. It was an innocent mistake, but one that Courts had to explain personally to the SEC for Delta to avoid any major penalties. This revealed that Monroe's lawyers, however well meaning, were out of their ken when it came to Delta's future.[83]

In addition, decisions made by the CAB further oriented Delta toward Atlanta. Beginning a strategy that would provide the direction for Delta's growth after World War II, Woolman tried to acquire a small company called "Marquette Air Lines," which served St. Louis, Missouri; Cincinnati, Ohio; and Detroit, Michigan. This attempted merger—along with Delta's requests to the CAB to serve Tallahassee, Jacksonville, and Tampa, Florida—had been refused. Finally, on January 30, 1941, the CAB allowed Delta to serve a new route north from Atlanta, Georgia to Knoxville, Tennessee; Lexington, Kentucky; and ending in Cincinnati, Ohio. With this development, Atlanta became the epicenter of Delta's progress from its past in the South toward its national future with an expansion toward the Midwest. Although many of Delta's stockholders resisted moving to Atlanta and favored relocating instead to Dallas (perhaps because that city was more familiar to Delta's investors from neighboring Louisiana), Delta's historians note that "the board . . . in this instance, as in others, . . . always followed Woolman's lead in the end."[84] After having been swayed by sentiment to stay in Monroe longer than was necessary, Woolman certainly was not going to make the same mistake twice by placating his stockholders' sentimental attachment to Dallas.

By 1940, Delta had become, to use a southern term that Woolman was probably familiar with, "too big for its britches." Although Woolman and his company could never have been accused of being arrogant (as the old saying implies), their growing pains were real. Difficult though it may have been to uproot themselves from familiar surroundings in Monroe, Woolman and other Delta employees were also savoring the fruits of their patience, frugality, and prudent risk-taking: in 1941, Delta flew 15,000,000 passenger miles, which helped to push its gross revenues above $1,000,000 for the first time.[85] Having survived so many setbacks in Delta's short history, Woolman was now steeled to take on the bigger airlines that had once threatened to put Delta out of business.

Challenging Eastern Airlines

Like Woolman, Eddie Rickenbacker made his mark in the South through his stewardship of Eastern Airlines. An extroverted and straight-talking war hero, Rickenbacker was admired by millions of Americans for his bravery. He considered himself a star and expected others to treat him as such. Demanding deference, even from government officials whose authority he was legally bound to obey, Rickenbacker's great energy in developing Eastern Airlines was often undercut by his inability to adjust to the changing political currents.

In contrast Woolman, who adapted to government bureaucracy, Rickenbacker prided himself on his company's separation from the U.S. government. According to one historian, Rickenbacker viewed success as "getting passengers where they wanted to go at reasonable prices that would yield profits to Eastern without federal subsidy." Resisting payments from the government "as a matter of principle," Rickenbacker put his company at a competitive disadvantage against airlines that were more than willing to take money from Uncle Sam.

Woolman and Delta were all too ready to partner with the government. Woolman's cooperation was evident during World War II when he placed the airline at the disposal of the U.S. government. During the war, all but four of Delta's planes and almost all of its pilots were deployed for military purposes.[86] Even before the end of the war, Delta reaped the benefits of its partnership with the government when it was allowed to provide airline services for the city of New Orleans. Shortly after the war, Woolman embarked on a major expansion plan for the airline, hoping to add a number of new routes that would allow the airline to offer passenger travel from the upper Midwestern states to Florida. In August 1945, Delta achieved this goal when the CAB authorized the airline to fly between Chicago and Miami. This 1,028 mile route was the longest single route that the CAB had awarded since its inception in 1938, and with this new route, Delta became a direct competitor of Eastern.[87]

Delta's expansion plans received a tremendous additional boost in 1953 when the CAB allowed the company to merge with the Chicago and Southern (C&S) Airline. In addition to several new domestic routes, C&S had been awarded a number of international routes to Caribbean and South American destinations through its hub in New Orleans.[88] After combining its operations with C&S, Delta also succeeded in its bid for the important 1,075 mile route from New York to Atlanta. This route became an even greater competitive threat to Eastern.

As Delta grew, it pioneered the "hub and spoke" model of airline operations whereby traffic from small and medium-sized cities was funneled to nodal points such as Atlanta, Memphis, and Chicago during specific time frames throughout the day. The traffic was then redistributed to longer-haul flights heading to major destination points. This model facilitated "an enormous variety of connections that would otherwise have been impossible...optimized load factors, conserved fuel, and encouraged Delta to hold onto smaller stations instead of turning them over to feeder lines."[89] The hub and spoke model that was originally pioneered in Atlanta by Delta became a key part of the business model for most traditional airlines during the postderegulation period of the airline industry. As Delta thrived in the postwar growth of the airline industry, Eastern struggled to maintain its hold on airline service. Delta's historians, Lewis and Newton note:

> In contending for potentially lucrative routes during this era, Delta did better than its major competitor, Eastern. This was in part because, under Woolman's leadership Delta made maximum use of its contacts with key southern representatives and senators at a time when the South was electing solidly Democratic delegations to Congress despite its infatuation at the national level with such Republican leaders as Dwight D. Eisenhower. Conversely, at the helm of Eastern, the politically archconservative Rickenbacker went out of his way to alienate influential Democrats and thus squandered some of his firm's potential influence in the regulatory arena.[90]

Calculated or not, Rickenbacker's violent disgust with government's role in shaping the postwar airline industry made it impossible for him to adjust to the changing times. While Eastern would cease to exist by the end of the twentieth century, Delta continued to expand. In many ways, Woolman made this possible through his transition from entrepreneur to manager. Similarly, C. R. Smith of American Airlines and Pat Patterson of United Air Lines, whose stories are told in the next section, were able to secure a dominant position for their airlines by deftly working within the parameters of government regulation that characterized the second phase of the airline industry's evolution.

PART II

The Managers

As we saw in Part I, various entrepreneurs competed to secure a sustainable business model in the airline industry. Although entrepreneurs shape the initial parameters of possible successful business models, it is often managers who maximize that potential through efforts to both reinforce and defend the dominant business model. In some cases, successful entrepreneurs become equally successful managers, helping to shepherd their company through growth and maturity with an eye toward the changing competitive landscape. In most cases, however, entrepreneurs struggle to make the leap from leading a start-up to managing an ongoing concern. Managers, who often have different skills and temperaments, step in to guide these businesses through growth and maturity.

To maximize growth potential, managers typically focus on standardization, economies of scale, differentiation, and efficiency. The early investments that are made by entrepreneurs and heavily supplemented by managers are designed to act as barriers to entry as well as opportunities to capitalize on the increasing scale and scope of industry development. The inherent success of a business model attracts several followers during the growth phase of an industry's lifecycle with each competitor vying to sustain success or to supplant another player. As an industry reaches maturity, managers tend to look for opportunities to sustain dominance through further efforts to gain scale. During the later stages of an industry's lifecycle, this is often accomplished by consolidations, mergers and acquisitions, or strategic alliances and partnerships.

The rapid expansion and consolidation of the airline industry can also be compared to similar forces in the automobile industry (albeit this was not an industry that was regulated). The figure that follows (figure M.1) on the global automobile industry in the twentieth century shows a period of drastic expansion sparked by entrepreneurs followed by massive consolidation orchestrated by managers.[1] At its peak in the 1910s, the global automobile industry was comprised of more than 500 manufacturers. Forty years later, the industry had

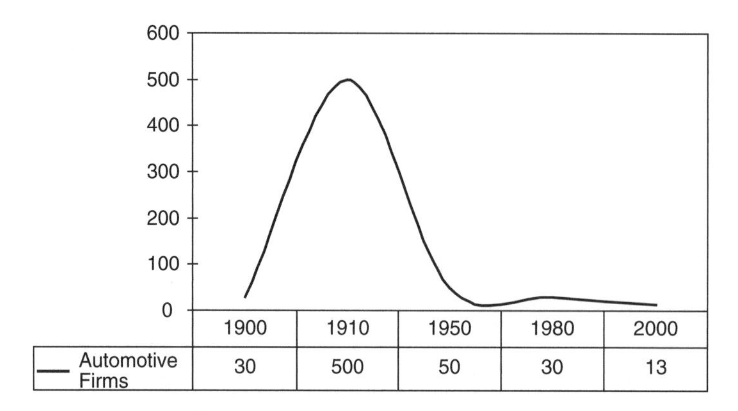

Figure M.1 Global automobile firms

shrunk to only 50 manufacturers. The airline industry encountered a similar phenomenon (see figure I.2 in the Introduction).

Before the mid-1930s, the airline industry was characterized by constant change, numerous start-ups, and diverse aircraft. With the aid of the U.S. government, the airline industry was drastically rationalized in the 1930s, and a dominant business model emerged led by American, United, TWA, and Eastern. Government officials believed that the future of the airline industry was predicated on maximizing scale and scope and, to that end, the country was divided into regional spheres with a dominant carrier at the helm—American flying the transcontinental southern route, United leading the transcontinental northern route, TWA dominating the transcontinental central route, and Eastern holding the leadership position in the heavily trafficked north-south corridor of the eastern seaboard. Juan Trippe and Pan Am also greatly benefited from this intervention as the U.S. government sought to create, defend, and reinforce a single international carrier for many years. Despite the concentration of activity within the airline industry, there were also a few pockets of opportunity for regional players like Delta Air Lines.

The formal regulation of the airline industry in 1938 solidified the dominant business model for four decades, and managers worked in concert with the government to defend their representative positions. Under regulation, airlines were restricted on the prices that could be charged to consumers, the routes that could be added or dropped, and the ability to collaborate with other carriers. Under these restrictions, managers of the major carriers sought ways to distinguish their airlines through the development of new consumer services and innovative technology. Woolman of Delta summed up the state of the industry during regulation when he said, "The airline industry is keenly competitive. All of us have good planes, the only way in which we can excel is in the quality of service. And this is where the human factor enters."[2] The "human factor" in the evolution of the airline industry was often supplied by the leader or leaders who headed the major airlines. It was the actions of these individuals who shaped the competitive framework of the industry within the parameters of regulation.

In many ways, managers acted in very innovative ways to differentiate their airlines, but more often than not, their successful experimentations were copied or adapted by other major carriers. In essence, managers' successful innovations became additional building blocks in the dominant business model. This elaboration of services, features, or enhancements to the dominant business model is typical of most industries as they mature. To retain a dominant position, competitors must minimally offer the same proposition to consumers even as the core of baseline services expands.

In Part II of this book, we will explore the role of managers who operated under the influence of government regulation. As with entrepreneurs, managers are not one-dimensional in their leadership approach and strategic choices. The three managers that we profile in this section of the book were chosen because they represent the three prototypical managerial archetypes–the *technology-focused manager*, the *market-focused manager*, and the *organizationally focused manager*. As Juan Trippe moved his company from the entrepreneurial to the managerial phase, he embodied the approach and tactics of the technology-focused manager. He used technology as a means of securing Pan Am's dominance in international travel. C. R. Smith of American Airlines is representative of the market-focused manager who concentrated on building a customer base by first addressing issues of safety and then rewarding loyal customers for their business. In addition, Smith used technology as a means of supporting his market-facing objectives. Finally, Pat Patterson of United Air Lines through his focus on operational efficiency and alignment is an exemplar of the organizationally focused manager. While all managers must be technology, market, and, organizationally focused to succeed, they tend to emphasize one area more than another.

In chapter 4, we will pick up the story of Juan Trippe and Pan American as the company comes to dominate the foreign airmail business and international passenger traffic. Through his deft efforts at lobbying, Trippe was able to secure Pan Am's dominance and exclusivity for several decades. He significantly reinforced Pan Am's ties to the government through the development of landing strips throughout the Pacific isles during World War II. Trippe also played an important role in influencing the technology of aircraft through his almost four decade tenure as the head of Pan Am. He funded the development of larger and more powerful aircraft that could traverse the great distances across the Pacific Ocean, and he was largely responsible for the introduction of jet aircraft in the 1960s and 1970s. Though he sometimes overextended the financial resources of his organization through his advocacy for more powerful aircraft, he forced the industry to move forward. To keep pace with and eventually compete head to head with Pan Am in the international segment of air travel, other carriers had no choice but to follow Trippe's technological lead.

C. R. Smith of American Airlines who we will meet in chapter 5 also played a role in shaping the technology of the airline industry. The 21-passenger DC-3 aircraft built by Douglas Aircraft that American Airlines put into service in the mid-1930s was the first commercially viable plane to transport passengers. Smith and his team worked closely with Donald Douglas on all design aspects of the DC-3. Smith also picked up the mantle of plane safety from Harry Guggenheim. Smith's introduction of the path-breaking "Afraid to Fly?" marketing campaign in 1937 attacked consumers' fears of flying with a direct, respectful, and informed perspective. While others chose to ignore or gloss over this issue, Smith believed that consumers were savvy enough to understand the low risks associated with air travel if they were presented with the facts in a compelling and straightforward manner. Smith's focus on safety was so persuasive that competitors simply copied his direct advertisements in their own outreach efforts.

Smith went on to introduce a number of other important consumer-oriented services such as the Admirals Club for frequent travelers and coach fares for families that were quickly adopted by others and infused into the dominant business model in the airline industry. Between the 1930s and late 1960s, the airline industry evolved through the reinforcing actions of airline CEOs like Smith, the U.S. government, and aircraft manufacturers. In many respects, Smith and American Airlines were at the forefront of this partnership. His efforts and those of his counterparts enabled the airline industry to expand through a defined set of provisions that favored consistency and standardization of services.

Pat Patterson of United whose story is told in chapter 6 also introduced a series of innovations including flight attendants and food on airlines that were swiftly adopted by others. In addition, Patterson built an organizational model of efficiency by centralizing technical and logistical resources in specific locations. He did so to effectively manage the scale and scope of United's rapidly expanding business. This organizational focus became a key part of building the infrastructure to support the largest airline in the United States. Under regulation, managers like Smith and Patterson needed to be creative to distinguish their businesses. Though both ran companies based on a dominant large-scale, transcontinental operation, they introduced new services and benefits to attract more passengers and to take advantage of the anticipated growth in passenger travel after World War II.

The regulatory environment of the airline industry reinforced the dominant business model for 40 years from 1938 to 1978. The relative stability during this period of time required managers to be innovative in the ways in which they differentiated their airlines, but it also contributed to a general level of stasis or complacency in the industry and may have contributed to the exceedingly long tenures of most of the CEOs of the major airlines and, consequently, their lack of effort to build a strong and effective pipeline of successive leaders. In addition to their own internal issues, most airlines were not adequately prepared for the changes in the competitive landscape that first percolated in the late 1960s and came to a head in the 1970s with the onset of deregulation. For 40 years, the managers of the major carriers worked to protect and defend a specific business model. Many of them were unable or unwilling to recognize the changes in the competitive landscape and this inability or blindness created an opportunity for others like Herb Kelleher of Southwest Airlines to develop a strong, alternative business model that was much better positioned for the contextual landscape of the airline industry at the end of the twentieth century. We will see the results of this overdependence on a single business model in Part III.

CHAPTER 4

Juan Trippe and the Growth of International Air Travel

Less than 10 years after Pan Am had to charter a plane to save itself from potential ruin by almost failing to fulfill its contract to deliver mail to Cuba, the company was playing a starring role in a major Hollywood movie.[1] *China Clipper* (1936)—starring, among others, Humphrey Bogart—portrayed the hard-driving Dave Logan, a pilot inspired to start his own airline after Lindbergh's solo flight across the Atlantic. After failing to launch a domestic airline that would serve Washington, D.C. and Philadelphia, Logan establishes an airline company to service the Caribbean and then sets his sights on an even bigger prize: transpacific air service from the United States to the Far East. His dreams are fulfilled by the use of a new, powerful airplane called, of course, the "China Clipper." The film was made with Juan Trippe's cooperation and eventually received his blessing. His associates at Pan Am were quite surprised that Trippe did not object to the final cut of the movie because the portrait of Dave Logan was not very flattering.[2] A film review in the *New York Times* provided an apt description of the character based on Pan Am's president: "he goads himself and his aides mercilessly toward the realization of his vision. Marriage, friendship, consideration of himself—these are sacrificed upon the altar of his ambition."[3]

China Clipper did not only reflect Trippe's colorful past, but it was also an important instrument of Trippe's future ambitions, for at the time the film was screened in August 1936, the real airline, Pan Am, was negotiating to obtain a goal that the fictitious airline had already acquired on the silver screen: the rights to land in China. The film ends with the airline reaching its destination at the Portuguese colony of Macao, located south of Shanghai on mainland China. Although Pan Am did not officially misrepresent that it had permits to fly only as far as the Philippines, the title of the movie, as well as the airplane it was named after, promised that airline service to China would be the company's main goal.[4]

Having built essentially a monopoly in Latin America, Trippe sought to protect and expand his franchise as the industry matured. He would, as he did 10 years before, often spread his resources too thin and push his airplanes and employees to the limit in an effort to meet the obligations to which he had committed the airline. His expert ability to influence U.S. government policy in the formative years of the development of international air travel enabled him to establish a dominant business model for this sector of the airline industry. Throughout the next three decades, Trippe worked to ensure that this dominant model was reinforced through continued government lobbying and international diplomacy as well as through technological leaps in aircraft production. Through these efforts Trippe was making the transition from entrepreneur to manager. Although he continued to innovate like many entrepreneurs, his efforts were squarely focused on protecting the monopolistic position he was able to forge for Pan Am. As he had done in the early phase of the airline industry, Trippe worked hard to align his goals with the goals of the U.S. government. This alignment was especially apparent in the years leading up to World War II; Pan Am functioned as a quasi-official strategic and diplomatic arm of the United States as it expanded its scope and reach throughout all corners of the world. As long as this alignment was maintained, Trippe and Pan Am were secure. While Trippe searched for ways to maintain his airline's dominant position in the international sector of the airline industry, the major domestic players, including C. R. Smith of American Airlines and Pat Patterson of United Air Lines (who we will see in chapters 5 and 6 respectively) along with C. E. Woolman of Delta Air Lines worked to secure and reinforce their positions within the domestic marketplace.

Securing Global Ambitions

After having secured international routes throughout the Caribbean and South America, the obvious next step for Trippe's Pan American would be to explore ways to offer service over the Atlantic. But Trippe, as his past actions had demonstrated, was not one to pursue the obvious and most cautious routes for his business. With Lindbergh acting as Pan Am's chief technical advisor, Trippe indulged his greatest aspirations by exploring Atlantic and Pacific routes *simultaneously*.

At this time, Lindbergh was eager to put his tragic past behind him—the awful days in the winter of 1932 when the world followed the kidnapping and murder of the aviator's infant son. After that, Lindbergh wanted nothing less than to fly around the world to evaluate possible routes for Pan Am. Lindbergh had already surveyed (at his own expense) a possible route from the United States to Asia in 1931. Because the maximum distance that aircraft could travel was approximately 1,000 miles at the time, the most promising route took the aviator north, toward Russia via the Bering Strait. Accompanied by his wife Anne, Lindbergh took his single-engine Lockheed Sirius seaplane all the way to China where he even helped local officials survey the immediate aftermath of a flood of the Yangtze River.

Despite his successful arrival in Asia, Lindbergh did not think that Pan American could make a northern route to the Pacific profitable. Too many problems dogged the aviator on his way to the East. Fog, frozen harbors, and unpredictable weather made air travel in this part of the world a constant challenge. To complicate matters, Lindbergh's progress was sometimes interfered with by some unknown electronic forces that weakened his radio signal intermittently.[5] At the end of their journey, the Lockheed Sirius overturned in the Yangtze River and photographs of the mishap were published the following month in the *New York Times*.[6] Besides the difficult terrain and weather that Pan American would have to surmount in establishing a route across the northern Pacific, the Russian government was reluctant to permit any U.S. business to work on their territory until the U.S. government officially recognized the communist regime. These obstacles notwithstanding, Trippe was still preparing the way for a northern route, buying two small airlines in Alaska for the relatively low price of $90,000 in the latter half of 1932.[7] Though the Northern Atlantic route was not feasible at this time, Trippe believed that Alaska would be a strategically important location in the future. Purchasing the only operating airline in Alaska, Trippe claimed this region for Pan Am at a price that was considered "dirt cheap."[8] In hindsight, the purchase was very astute. As technology and diplomatic relations improved, Alaska would become a linchpin in Pan Am's Asia- Pacific market. Managers like Trippe must constantly scour the horizon for competitive threats and opportunities, weigh various options, and hedge their bets to secure their dominance in the marketplace. For Trippe, the Alaska airline purchase was a relatively easy and low risk bet to make.

With difficulties in establishing a northern route to Asia in the early 1930s, Trippe began what would become very long and belabored negotiations with European governments and airlines to forge a transatlantic route. In 1933, Lindbergh surveyed some possible routes to Europe and found that the way across the northern Atlantic was just as inhospitable as the northern Pacific route. Instead, Lindbergh counseled Trippe to go to Europe over the central Atlantic Ocean via Bermuda and the Azores. Although logistical problems in crossing the mid-Atlantic would be difficult to overcome (the small harbors in the Azores were less than ideal landing spots for seaplanes), other obstacles would scuttle Trippe's plans to link the Old World with the New. Specifically, Trippe had to break off negotiations with the British when they demanded that any concessions given to Pan American were to be reciprocated by the U.S. government to British airlines interested in providing their own Atlantic service.[9] Trippe needed British cooperation because all routes to Europe required passage through some part of the British Commonwealth. As a private citizen, albeit well connected in the U.S. government, Trippe was hardly in a position to grant these requests.

Anticipating smooth sailing over the diplomatic waters of the Atlantic, Trippe had already committed Pan Am to buying approximately $2 million worth of seaplanes from the Martin and Sikorsky companies. Options were limited—the least attractive of which was canceling the order, a drastic measure that would offend airline manufacturers and make Pan Am seem like a company on the wane.[10] For Trippe, retreat was not an option in the fall of 1933. The company would have to continue to grow, and the only direction it could grow was toward the Pacific.

Three major problems needed to be solved before Pan Am could connect East to West. First, finding viable harbors to provision and refuel seaplanes would be difficult. The only possible harbors that were visible on the globe looked like tiny grains of sand scattered randomly across the map of the Pacific. Second, acquiring landing rights on the mainland of China ran into the obstacle of Chinese fears that their arch rivals, the Japanese, would demand landing rights in China if an American company were allowed to land on Chinese territory.[11] Third, Pan Am required seaplanes that were sturdy enough to weather repeated crossings of the vast ocean and comfortable enough to attract the wealthy clientele who could afford the costs of inaugurating transpacific air service. Solving the problem of developing needed airplane technology would inspire Trippe to come up with a novel way of accelerating the pace of innovation by airplane manufacturers, a topic we will take up in more detail later in this chapter. For the first two problems, Trippe again leveraged his airline's capacity to resolve sticky foreign policy conundrums faced by the U.S. government, even when the Roosevelt administration was dominated by people who strongly opposed government-sanctioned monopolies.

Luckily for Trippe, the U.S. government needed access to the Pacific as much as Pan American did. While Trippe was searching for ways to cross that 9,000-mile wide route in 1934, the United States was trying its best to respond to Japanese ambitions to expand its power and influence across Asia. Japanese-American diplomatic relations in the Pacific had been framed by the Five-Power Treaty of 1922, which involved Great Britain, the United States, Japan, France, and Italy; the signatories agreed to a ratio of capital ships (i.e., the most costly and powerful ships belonging to a navy) of 5 : 5 : 3 : 1.67 : 1.67, respectively. For the Harding administration, these talks presented an opportunity to forestall an arms race in the Pacific; these savings and the more peaceful diplomatic atmosphere they might engender would help the U.S. government to save money as well as to spur worldwide economic development in the wake of the recession of 1920–1921.[12] By the 1930s, however, the United States had become increasingly dependent on the treaty as a way to keep defense costs low during the cash-strapped days of the Great Depression.

Although the United States wanted to keep military tensions low in the 1930s, the Japanese Navy became especially aggressive in asserting its regional ambitions. As historian Akira Iriye explains, the Five-Power Treaty and others like it "were considered inimical to Japanese security, as they established an inferior naval ratio for Japan vis-à-vis the United States. It became the Japanese naval strategists' main concern to establish parity with American fleet strength and to prepare for southern expansion to control key areas in south China and southeast Asia."[13] Japan was likely further encouraged by a deal struck by the Philippine government with President Franklin Roosevelt to remove all American military installations from the Philippine mainland and to return soon to the negotiation table to discuss the fate of U.S. naval bases.[14] Not fearful that the United States would venture to enforce any of its policy goals through military actions, Japan grew increasingly bold and informed the U.S. government in September 1934 that it was planning to abrogate the Five-Power Treaty "at the earliest date possible, two years after December 31, 1934."[15] With this notice, the Roosevelt administration did not want to stand by passively and allow the next 27 months to elapse without any response. But any possible short-term reaction would have to be very carefully implemented, for the United States desperately wanted the Japanese to adhere to their treaty obligations.

Enter Juan Trippe and Pan American airlines. By October 1934, just after the Japanese announced their plan to increase their naval armaments, Trippe informed the U.S. Navy and the postmaster general that Pan Am was making plans for transpacific air service and would need to have access to marine airports at Guam, Wake, and Midway Islands (airports that did not yet exist). The timing of Trippe's message could hardly have been coincidental. Soon afterward, the State Department informed the navy that developing these islands for potential military use was a priority, but that it had to be done with the lightest of touches, lest the Japanese interpret these moves as aggressive steps by the United States. Without any fanfare, Roosevelt transferred many of the islands Pan Am was hoping to use for its seaplanes to the jurisdiction of the navy on December 13, 1934. Trippe had just smoothed the way for Pan Am's Pacific ambitions during the previous week by implying in a message to the secretary of the navy that the airline could act as a "surrogate" of the navy, by transforming tiny, isolated, and uninhabited islands into airports with infrastructure suitable for seaplanes, Trippe would be making them suitable for naval uses as well. Within a year, Pan Am had received the navy's permission to fly all the way to the Philippines.[16]

The skill of Pan Am's president in using his government contacts to promote his company's interests in the Pacific was all the more remarkable because of the radical change in government administration from the Hoover to the Roosevelt regimes. Although Trippe himself leaned Republican, his company would have no political affiliation. His nonpartisan lobbying efforts began early in 1933, opening the doors to opportunities in the Pacific.[17] These efforts also helped him to avoid receiving too much scrutiny from the Senator Hugo Black committee hearings of 1934, which had ruined the careers of many airline executives who had run the industry with the cooperation of Postmaster Walter Folger Brown in the early 1930s.

With this window of opportunity nudged open by Pan Am's relationship with the U.S. government, Trippe again took the initiative and forced open that window as far as he could. Before the navy had granted formal permission to construct facilities that would support seaplane operations, Trippe began the construction project on his own. Local investors in Hawaii were courted by Trippe and succumbed to his persuasive powers, buying $1 million dollars in Pan Am stock and promising to provide useful services to Pan Am seaplanes, such as weather reports. Trippe then arranged to transport more than 100 men and $500,000 worth of cargo to begin construction on Midway Island in the summer of 1935. The navy, of course, disavowed any official connection to the project; many of the navy's own personnel, convinced that Midway was a navy project, were disappointed to find that their applications to volunteer for this adventurous service were rejected. But behind the scenes, the navy was following

Trippe's lead and quietly aided in secretly discharging navy personnel who would then assume the guise of civilians employed to install and operate the radio communications that the China Clippers would need. Midway would later serve as a base for naval war games.[18]

After receiving *carte blanche* to fly all the way to the Philippines, Pan Am still needed to complete the last leg of its planned route to China. This would require Trippe's skills at exploiting international economic and diplomatic rivalries. The way had already been prepared for Pan Am months before by an equally adept emissary of Trippe's, Harold Bixby, a former banker from St. Louis who arrived in China in 1932. In March 1933, Trippe bought the American-run China National Aviation Corporation (CNAC), an airline serving large cities on China's coast that was on the verge of bankruptcy. The nationalistic Chinese balked at another American corporation buying what, in name at least, seemed to be a Chinese company. Bixby coaxed the Chinese to reluctantly agree by threatening to move CNAC's headquarters to the British colony of Hong Kong.[19]

While Bixby was keeping a small but firm toehold for Pan Am's operations in mainland China, Trippe angled for the larger prize of Hong Kong. The "China Clipper" airmail service had already begun in November 1935, but it went only as far as Manila.[20] While the movie *China Clipper* was raising the interest among would-be passengers to fly to China in the summer of 1936, Trippe was using the movie's fictitious ending—the "China Clipper" landing in Macao—as a means to pressure the British to give up their opposition to his company's entrance into Hong Kong. In fact, the movie depicted exactly what Trippe was doing. With life imitating fiction, Pan Am signed a five-year agreement that very summer with Portugal and its colony to base Pan Am's Chinese operations in Macao. Hong Kong's officials now began to lose their complaisance about their port facilities, which were far superior to those of their Portuguese competitors. Trippe turned Hong Kong's smugness to fear when he turned down a polite invitation from the Hong Kong government to visit their colony. The shoe was now on the other foot. Would the British be shut out of the emerging field of international aviation in the Pacific? Now believing that Trippe was not at all bluffing, businessmen on the island pressured British officials to make sure that Pan Am would be welcome in Hong Kong. The British government was clearly outmaneuvered. After having jealously guarded the privilege of landing in Hong Kong, the British turned 180 degrees and made Hong Kong "a free international airport" open to all nations without demands for reciprocal agreements, as had long been the British policy. By September, Pan Am and CNAC received the rights to land in Hong Kong.[21] In just five short years after having secured its Latin American routes, a Pan Am passenger could now travel from Buenos Aires to Hong Kong on a single airline.

Technology Race

To traverse such great distances, Trippe accelerated the process of technical innovation in ocean flight. At the time, Russian immigrant Igor Sikorsky was a leading engineer in the aircraft production industry. Sikorsky decided to leave for the United States during the Russian Revolution and eventually found enough investors to begin his own company, the Sikorsky Aero Engineering Corporation. Sikorsky's "amphibian" planes (amphibian because they were able to land on water and land) were used during Pan American's early days in surveys of the Caribbean and eventually became an important part of Pan Am's fleet when the company acquired nine Sikorsky "S-38" seaplanes in its purchase of NYRBA.[22]

These S-38s were small (room for 7 passengers and 3 crew members) and slow (cruising speed of 100 mph), but they were the most efficient planes available at the time. The S-38s, however, were insufficient for Trippe's aggressive ambitions; he commissioned a better airplane from Sikorsky. Pan Am's position as the only major U.S. carrier flying over the ocean gave it considerable influence over companies such as the Sikorsky Corporation that needed to sell seaplanes. Pan American's precious business came with a stipulation Sikorsky could not refuse: constant supervision and critiques from many of Pan American's executives, especially the company's lead technical consultant, Charles Lindbergh.

Trippe decided to try to accelerate the process of technical innovations in ocean flight by placing bids for three planes at two companies: the Sikorsky Corporation and another company run by Glen Martin who had made his name making large planes for the military. Trippe heightened the sense of competition by making a very public announcement of his bids at a press conference where he was accompanied by Lindbergh.[23] The technical achievements resulting from this competition were spectacular. The Sikorsky's S-42, whose first flight was in 1934, would eventually "set ten world records for seaplane performance" and could carry 32 passengers on a 1,000-mile journey nonstop cruising at 150 mph. The amount of weight its wings could carry "was 28.5 pounds per square foot, which was mind-boggling in an airliner of this date."[24] Nine months later, Martin's larger M-130 was introduced to the public. In the early stages of Martin's relationship with Pan Am, he benefited greatly from contact with Pan Am, for Trippe—putting his impatience to grow his company ahead of honor—allowed the lessons learned from Sikorsky's mistakes to be quietly filtered by Pan Am's technical advisors to the Martin manufacturing plant.[25]

When Trippe turned his sights on crossing the Pacific Ocean, he needed once again to improve airplane performance because simply shifting the M-130 to work in the Pacific Ocean would not bring about the necessary results. The Martin aircraft was designed to carry 46 passengers over the longest gap in the route over the North Atlantic, which was 2,000 miles between Newfoundland and Ireland. In a flight over the Pacific, the longest distance between ports would be the 2,410 miles of ocean spanning the gap between California and Hawaii. At first, Trippe did try to use the M-130 in the Pacific and advertised the Martin as a "forty-six-passenger airliner," but the extra fuel required to stretch the plane's range an extra 20 percent meant that it "rarely carried a dozen passengers."[26]

Glen Martin, who invested heavily in the prospects of a long-term relationship with Pan American, was outraged when, in 1936, he heard that Trippe sent out a call for airplane manufacturers to create seaplanes that could cruise 4,800 miles "loaded with 8,000 pounds of mail and cargo, and with accommodations for 50 passengers 'equal to the best available in rail transport.'"[27] After looking over Martin's proposals for redesigning the M-130, Pan Am executives rejected the plane as inadequate—too few passengers and too expensive. Sikorsky's design for the S-45 model, Lindbergh's choice, was rejected because it would take four years to develop. Trippe settled on the design submitted by Boeing for its B-314, which was promised for delivery in September 1937 (but which eventually finished in January 1939). Although less powerful than the Sikorsky model, the Boeing represented yet another quantum leap in airplane development, boasting a maximum takeoff weight of 82,500 pounds, which represented enough power to take 21 passengers and 8 crew members 2,500 miles.[28] Forty years later, Trippe recalled the reasons why Pan Am abandoned Martin for Boeing: "Sure, Martin lost money, but he didn't have the next step."[29] By forcing various aircraft manufacturers to compete for Pan Am's business, Trippe fundamentally shaped the technology that was both achievable and ultimately necessary for international carriers. Through his efforts, Trippe was often able to obtain the best and most powerful aircraft years ahead of others, which further reinforced the dominance of Pan Am's business model.

Although completing the trip between San Francisco and Hong Kong was a logistical and public relations triumph, the real money to be made in international flights was over the Atlantic, where historic and economic ties to the United States were strongest and logistical problems were less pronounced than over the Pacific. For instance, Chinese economic policies would shut down Pan Am's operations in China completely by 1949.[30] Trippe's success in the Pacific in 1935 was

The giant 74-passenger Yankee Clipper, pictured flying over the Capitol Building en route to the Naval Air Station, Anacostia, where it was christened by Mrs. Roosevelt. (Source: Bettmann/CORBIS).

complemented four years later when Pan Am's Yankee Clipper (a Boeing B-314) made the first scheduled transatlantic flight from New York to France. The first transatlantic passenger flight took 22 hours and cost $375 each way.[31] The 22 passengers on the first flight enjoyed a 6-course dinner and 5-course breakfast.[32]

Trippe Deposed

When Pan Am made its inaugural flight over the Atlantic in June 1939, Trippe was no longer the chief executive officer of the company. A few months before the historic flight, Trippe had been stripped of much of his power and replaced by Sonny Whitney, his lifelong friend and benefactor (Whitney had invested in both Colonial Air Transport and Pan American). At the time of the transition of power, Whitney had been serving as the chairman of Pan Am's board of directors. In early 1939, the board had become increasingly concerned with Trippe's expansionist vision and his reclusiveness. The decision to pursue air travel over the Atlantic and Pacific Oceans simultaneously stretched the resources of the company to the point where Pan Am had fully extended its financial credit lines and was unable to pay its shareholders dividends. By the beginning of 1939, Pan Am's operations spanned 54,072 route miles in 47 countries.[33] In addition to the stretched finances, the board was concerned with the overall management of the firm. Harold Bixby, Trippe's envoy in the negotiations with China, characterized the mood of the company as follows in a 1939 note to Lindbergh.

> Some nine months in the NY office has convinced me that-
>
> 1. J. T. T[rippe] will *never* delegate authority—with the result that he is partial to "yes" men—a bad omen for me because I will not conform.
> 2. Paper work—the servant—has become the master in the PAA organization and most personnel are so busy writing about their work that they have little time left to do it.
> 3. The human side of personal contacts and the inspiration of real leadership [have] been lost—displaced by bulletins and circulars. The executives in NY write letters—instead of going over [to] their divisions.[34]

As the organization grew, Trippe held onto his autocratic and controlling leadership style. While that leadership style can be appropriate at certain inflection points in a company's evolution (during start-up phases or in times of turmoil), it does not always serve the company well as the business matures and business lines expand. Under these growth circumstances, effective managers must delegate and communicate more. Trippe struggled with the move from entrepreneur to manager in his leadership style with subordinates and with the board. He still wanted to control all aspects of the organization, which was consistent with his earlier leadership approach at Colonial Air Transport. Trippe was, however, adept as a manager in his ability to secure and maintain Pan Am's dominance in foreign airmail and international passenger traffic. His downfall was that he pursued this path without seeking advice from the board or counsel from others within the organization.

As the largest shareholder in the company, Whitney helped to orchestrate the coup d'etat by calling a special board meeting on March 14, 1939. The board voted to amend the company's bylaws to name the chairman of the board the new chief executive and to allow Trippe to stay on as president and general manager.[35] The board was most concerned about Trippe's failure to communicate his plans for the company. Despite the change in roles, Trippe maintained a tight sense of control over the organization. Trippe's biographers Marilyn Bender and Selig Altschul note:

> Sonny held weekly meetings on Wednesdays at 9 a.m. He went around the table asking each executive to report on his phase of operation. Trippe was a silent hulk, exuding fury. The others sensed he was rating them on the degree of their cooperativeness towards Whitney, storing their grades away in the recesses of his mind, biding his time toward a day of reckoning.[36]

The company's financial situation took a positive turn in September 1939 when the foreign airmail rates paid by the government were raised from $2/mile to $3.35/mile. The U.S. Navy was a key contributor to supporting the increase in mail rates by testifying before the Civil Aeronautics Board (CAB) that "Pan American had generously put its bases at the navy's disposition; its navigational and weather-reporting systems were useful to the defense of the Pacific and to shielding Hawaii from enemy attack."[37] The rate increase provided significant financial relief for the company, but despite the reversal of fortunes, Whitney struggled to bring the operation under control. Bender and Altschul recounted:

> It had been discovered that the keys to the Pan American Airways system were locked in Juan Trippe's head, and true to his old nickname, Mummy was not about to hand them over. Problems kept popping up for which the files yielded no answers, and there was no use asking other executives. Some of the commitments Juan had made seemed to be bound up with national security, and no one knew how to go about unraveling them, or whether, indeed, they should be let alone.[38]

The board became increasingly concerned about Whitney's lack of knowledge about the company's affairs. Nine months after he had been stripped of his post, Trippe was reinstated as the chief executive by the board. Whitney had long grown tired of trying to wrestle control from Trippe and decided to support the transition. In stepping aside, Whitney also walked away from the company, selling off most of his holdings in 1940 and 1941. One of the conditions of Trippe's reinstatement was the appointment of an administrative vice president to oversee operations and provide regular reports to the board. Though Trippe bristled at the appointment, his acceptance signaled an important shift in his role as a manager. One way in which entrepreneurs, who struggle to lead their companies as they mature, make the transition to manager is to surround themselves with individuals who can complement their leadership style and approach and who can pursue the operational and administrative aspects of the business that become more important as the company evolves. This co-leadership approach enables the entrepreneur to focus on the areas of the business that energize him without ignoring critical operational needs.

Pan American during the War Years

Trippe regained the helm of Pan American at a time when the company was increasingly called upon to support the strategic military initiatives of the United States in the Pacific region. The development of the Clipper Aircraft (known as the "flying boat") for Pan Am's global operations became an important component of the U.S. military effort during World War II. With their giant capacity for hauling freight and passengers, the Clippers were fully deployed as military transport carriers, often flying 12 hours a day.[39] Although most domestic carriers were able to retain a significant portion of their commercial operations, more than half of Pan Am's routes were within war zones, which required almost immediate conversion for military purposes.[40] Although Pan Am's equipment was deployed during the war, Trippe's personal adjustment to wartime needs was mixed. While he helped the Roosevelt administration develop airports friendly to U.S. interests in South America, he refused a War Department request from Hap Arnold to enlist officially in the army as a brigadier general to run the Air Transport Command.[41] This role was eventually assumed by C. R. Smith of American Airlines. Trippe bristled at the idea of being part of a formal bureaucracy in which he was not the ultimate leader. He commented that "he was too accustomed to being the final authority to fit into a military organization."[42] Trippe ultimately believed he could do more for the country at the helm of Pam Am. In the end, Trippe may have hurt Pan Am's attempt to maintain its U.S. monopoly over international travel by this uncharacteristic misstep in his relations with government officials.

As historian Wesley Newton notes, General Hap Arnold was one of the original founders of Pan American Airways, a company that he had seen slip from his fingers through the artful manipulation of Trippe and Hoyt (see chapter 2 on Pan Am for more details). Just after Trippe's refusal, "with the strong encouragement of Arnold, 21 other American airlines began to carry military cargo to various areas of the world." Arnold was more explicit "in a meeting he called in 1943 of U.S. airline representatives, including Pan American, at which he urged them (ordered, by some accounts) to expect worldwide competition in postwar skies for the health of future American Air Power."[43]

Despite the long-term implications for Pan Am, the war provided some unexpected immediate benefits for the company. Between 1939 and 1941 Pan American's route system grew to 98,582 miles, which was larger than the combined route total of all European carriers.[44] During the same time period, Pan Am's earnings increased by 76 percent.[45] Much of the increase in revenues and earnings stemmed from contracts to build 50 new airports throughout the Pacific region. Pan Am's domination of international travel was so strong that one official from the Central Intelligence Agency noted: "Pan Am was the American flag, for all practical purposes, an extension of the

United States Government. In many places, it was the only symbol of America besides the embassy. Even before there was Coca-Cola in some places, there was Pan Am."[46]

The End of the Chosen Instrument

When World War II drew to a close, so too did the U.S. government's favorable treatment of Pan Am. Even before the war ended, several domestic airlines encouraged by government officials, including Hap Arnold, announced that after the war they would file international route applications with the CAB, the government body charged with assigning new air routes and rate schedules. At a joint conference to review their postwar options for international expansion, the domestic airlines issued the following policy statement: "there can be no rational basis for permitting air transport outside the United States to be 'left to the withering influence of monopoly.' "[47] Trying to stem the rising tide of international competition, Trippe offered to make Pan Am a regulated monopoly. As part of his *Plan for the Consolidation of All American-Flag Overseas and Foreign Air Transport Operations*, Trippe noted:

> Every other major trading nation concentrated its strength behind a single "chosen instrument" airline. America in the postwar must do the same. American companies competing against each other in foreign countries would be obliged to vie for the favor of the foreign governments to the detriment of their own best interests. The U.S. government would not be able to support any one of them. The foreigners, meanwhile, would be heavily subsidized by their own governments, and would engage in practices inimical to the Americans. For instance, the foreigners would feed all traffic generated in their countries into their own airlines. In America, the many airlines were at odds, and they would feed their traffic into whatever foreign-flag airlines fed them back, not caring how this hurt competing American-flag airlines, and not caring either how much it might add to America's balance of payments deficits.[48]

At the time of his proposal, there were only three U.S.-based airlines operating regularly scheduled commercial flights outside the United States—Pan American, Panagra (partially owned by Pan Am), and American Export. Trippe's plan called for the consolidation of the three airlines into a new airline that he called a Community Company. The Community Company would issue $200 million worth of common stock, with one quarter being allocated to the three component airlines in proportion to their actual assets and the balance (with proportional distribution to be determined) to the rest of the American transportation industry, including the domestic airlines, railroads, shipping lines, and bus lines.[49] Although not explicit in the plan, as the largest international carrier, Pan American and its officials would be in the best position to oversee the new Community Company.

Trippe's plan for the Community Company formed the basis of the "All American-Flag Line" bill that was introduced in the U.S. Senate in 1944. Trippe ardently defended the bill in a series of public hearings and testimony over the next

Juan T. Trippe testifying before the U.S. Senate. (Source: Bettmann/CORBIS).

year, but he was unable to overcome the many constituents who opposed the bill. Representatives from the State, War, Navy and Justice Departments all testified against the bill. At the same time that the "All American-Flag Line" bill was making its way through the Senate's Commerce Committee, the CAB was reviewing applications from Transcontinental and Western Air (TWA), Pan American, and American Export for international routes over the North Atlantic. The CAB ruled on July 5, 1945, to grant new international routes to all three applicants. The Commerce Committee's recommendation to the Senate for the "All American-Flag Line" bill was scheduled for the following day, July 6, 1945, but "when the commerce committee sat down on July 6 to vote on the Community Company bill, many senators felt that public policy had already been decided. The CAB ruling had made [the] bill moot. A vote was taken nonetheless. The result was ten to ten... The tie vote meant that [the bill] would not be reported out on the Senate floor for debate and possible enactment into law."[50]

With the defeat of the Community Company bill, competition for the international passenger intensified. By the end of the 1960s, the number of U.S.-based carriers serving the California to Hawaii (see sidebar)[51] route increased from two to seven, the number serving the Far East increased from two to four, and in the South Pacific, Pan Am faced one additional competitor.[52] Pan Am's fate was further complicated by the CAB after it approved TWA's request to significantly expand its routes to Europe. Making matters worse, Pan Am had not secured any domestic routes. Trippe's deep desire to protect his international business actually undercut the airline's ability to survive in an evolving competitive landscape. His sole dedication to trying to defend his company's near monopoly on international travel from the United States seemingly prevented him from pursuing domestic airline routes that could have been strong feeder systems for Pan Am's broader international system. The seeds of the company's ultimate destruction were sown in Trippe's singular focus on international travel.

Donald Nyrop and Juan Trippe: Battle of the Diplomats

Shortly after Nyrop took over the leadership of Northwest in 1954, the airline's presumptive hold on Asia was challenged by none other than Trippe. Northwest's service to Asia was a centerpiece of its long-term business strategy, and taking that away would be fatal to the airline's future. Northwest's strategy, however, included a weak link: its transpacific routes were quickly approved by the CAB in 1946 on a temporary basis. Northwest also shared the very profitable route of Seattle-Portland to Hawaii with its rival Pan American. Moreover, that route was also awarded by the CAB in 1948 on a temporary basis. By 1955, it was time to renew the contracts.

Trippe's Pan American, with its extensive and deep contacts to the Washington establishment, struck hard and fast to lay an exclusive claim to some of Northwest's vital links to the East. At first, the CAB had decided unanimously in early January 1955 that the Seattle-Portland to Hawaii route should be run exclusively by Northwest. But just a couple of weeks later, Trippe's lobbyists targeted their efforts at persuading the executive branch that the subsidies Pan American required from the government were lower than those of Northwest. That argument struck a chord with President Dwight D. Eisenhower, who was eager to cut the government's financial support of airlines. In February, Eisenhower surprised the airline industry by reversing the decision of the CAB. Pan American was now given the Seattle-Portland to Hawaii route; in addition, Eisenhower left the door open for Pan American to compete directly with (or perhaps even replace) Northwest on its "great circle" route to Asia via Alaska.

Nyrop reacted immediately by traveling straight to Washington himself. Unlike Trippe, Nyrop had not sought to gain advantage in Washington by putting lobbyists on the company payroll. Instead, in this instance, he brought together his deep connections to the federal bureaucracy overseeing the airline industry (he was head of the CAB from 1951 to 1952) along with his thorough understanding of his company's finances to counter this impressive thrust by Pan American into the heart of Northwest's operations.

First, he contacted the Republican Senator from Minnesota Edward Thye as well as Congressman Walter Judd. They, in turn, got the attention of the president by calling a Saturday meeting at the White House. Second, he supplied those representatives from Minnesota with the facts that might persuade the president to change his mind.

Eisenhower was annoyed by the Saturday visit, reportedly pounding his fists on his desk in frustration with attending this unusual gathering. But the

Minnesota delegation patiently demonstrated that Pan American's description of Northwest's needs for government subsidies were based on incomplete information. Although it was true that during the two years before Nyrop became president Northwest lagged behind Pan American in passengers and revenues to Hawaii, the airline was quickly turning that situation around in 1954.

Indeed, since 1948 when Pan American and Northwest began to serve Hawaii on the same route, Northwest had actually carried more passengers: 31,038 compared to Pan Am's 30,700. Most importantly, Nyrop's numbers showed that Northwest could do what he had recently promised in a public statement: to fly passengers to Hawaii without government subsidy by 1956. Eisenhower was embarrassed by his decision to overturn a unanimous decision by the CAB because of faulty information. He quickly reversed his decision, gave Northwest the Hawaii route for three more years (in competition with Pan Am), and soon afterward admitted in a press conference that he had "made an error." Although Trippe's influence in Washington would remain a significant problem for Northwest's Asian ambitions in the following years, Nyrop's victory in 1956 kept Northwest's fortunes favorable. The possibility of losing his company's business in Asia alerted Nyrop to the need of expanding Northwest's domestic route system, which became a major priority for the airline. Unfortunately for Pan Am, Trippe did not feel the same level of anxiety.

Intensified Competition

In an effort to grow his international passenger service, Trippe often pushed for lower fares as soon as he could. Although the fares for connecting the United States to Asia would not drop significantly between the 1930s and the early 1950s—due mainly to the fact that demand was relatively stable—Trippe pushed down fares within the United States and the Western Hemisphere and kept them low when demand rose in response to cheaper tickets.[53] Round-trip fares between San Francisco and Hawaii dropped from $648 in 1936 to $173 dollars in 1944.[54]

Reductions of prices for travel in the Caribbean and South America were even more dramatic. Trippe introduced what would soon be widely known as the economical "tourist fare" on the New York to San Juan route, reducing one-way tickets from $131 to $75 in September 1948 for flights in modified DC-4s whose seating capacity was increased from 44 to 61 passengers. The *New York Times* marveled, remarking that the "round-trip fare of $150 is, incidentally, the lowest in the world for such an ocean hop."[55] Response was overwhelming—so much so that, according to two of Trippe's biographers, Pan Am's fares could be seen as the main engine behind the growth of the Puerto Rican population in New York City.[56] Although Pan American needed only to receive approval from the United States' CAB to implement these tourist-class fares, there were other problems on the horizon, including new bureaucratic hurdles to overcome in order to introduce the same kind of service to Europe as well as a new era of competition at home and abroad.

A few months before the defeat of the Community Company bill, airline representatives from 25 nations met in Havana, Cuba to form the International Air Transport Association (IATA) to better manage international airline traffic, including setting rates for passenger fares. Receiving approval from their respective governments, who were eager to accelerate tourism and commerce between America and Europe, the mainly European members of the IATA essentially became the governing body of international aviation because their recommendations were almost always enacted into law. Before the first working meeting of the organization, held in Montreal during October 1945, the European members of the IATA had come to an important consensus: one-way fares between New York and London should be set at $572.[57]

This startlingly high fare—prewar one-way tickets cost $375—proved to be a red flag for Trippe, who reacted with his characteristic force and audacity. Pan American bluntly announced at the conference that it intended to offer something even lower than prewar prices: a one-way North Atlantic fare of $275![58] Although Trippe made this proposal largely for the benefit of the U.S. government (demonstrating that Trippe's still-fragile monopoly would not stand in the way of low prices), this proposal was no fluke.[59] Over the next four years, Trippe provoked and badgered his peers at the IATA to approve his tourist fares to Europe. Getting close to this elusive goal in the summer of 1951, Trippe reached a compromise with other members of the IATA only to be thwarted by the U.S. CAB; the latter objected to the stipulation that there should be "frequency limitations" of transatlantic flights with tourist-class fares.[60]

This issue came to a head in the fall of 1951. Overall, transatlantic travel was stagnating for U.S. airlines since the end of World War II. Even worse, U.S. airlines were taking home much less of the transatlantic pie, dropping from carrying 83 percent of the transatlantic market in 1946 to just 57 percent in 1951. With European members of the IATA obviously benefiting from their stonewalling of Trippe's low-cost proposals, Trippe decided to take the gloves off. In November 1951, Trippe threatened to end Pan Am's association with the IATA unless it agreed to his proposals. Postwar Europe, still suffering from the destruction of World War II, needed as much economic cooperation with the United States as possible. Furthermore, Trippe's vocal denunciations of the IATA presented a public-relations nightmare for European airlines that did not want to be portrayed as obstructing lower cost fares for the common man. On November 27, 1951, Trippe finally succeeded in introducing low-cost transatlantic fares to the traveling public that would go into effect on April 1 of the following year.[61]

The results for Pan Am were heartening. According to the company's annual reports, net income for the company jumped from $2.4 million dollars in 1949 to $10.8 million in 1953. Trippe's triumph was further solidified after two more major accomplishments. In September 1954, Trippe was named president of the IATA after having been its nemesis since its inception.[62] And, quietly but perhaps even more significantly, Pan Am also noted its role in setting a major milestone for the traveling public: "In 1954, the airplane replaced the surface vessel as the principle medium for overseas travel. The number of passengers who preferred sea transport did not decrease. For the first time, however, those who selected air travel were in the majority." Furthermore as noted in Pan Am's annual report, "in 1954, 28% of all travelers going overseas to and from the continental United States, whether by sea or by air, chose to go by Pan American Clipper."[63]

The Jet Age

To support the increase in international air travel, the jet airliner became the centerpiece of the airline industry in the mid-1950s. Reversing Britain's sluggishness behind the American seaplane industry in the 1930s, the British De Havilland Aircraft Company raced ahead of the competition with the Comet in 1952, the world's first jet airliner. One early passenger of the Comet in 1953 observed that while the "turbo-prop is less noisy than the piston-engine airplane, the pure jet is even more quiet. Both new planes have less vibration than the old-fashioned planes, the jet practically none."[64] This time, it was the Americans who were skeptical of British technological advances and British claims "that the Comet is highly popular with the public and an economical plane."[65] Although many American airline CEOs, such as American's C.R.Smith, protested that the cost of buying and maintaining jets made their economic viability suspect in the short-term, Trippe hedged his bets, ordering three advanced-model Comets for Pan Am, scheduled for delivery in 1956. With Pan Am just having overcome the IATA's obstruction of tourist-class fares, jet airliners with larger passenger capacity than the Comet promised economies of scale that could dwarf those of standard propeller-driven aircraft.[66] Trippe firmly believed that jets were part of his company's immediate future, remarking: "If business dictates[,] Pan American will buy British aircraft."[67]

The British head start in manufacturing jet aircraft was undercut by the mysterious crashes of three Comets between January and April 1954. The planes had fallen from the sky for no immediately apparent reason. Likening the scientific inquiry into the Comet's demise to a crime-scene investigation, the *New York Times* reported that the researchers "at Farnborough [England] think they know what killed her. The murderer was metal fatigue."[68]

Britain's great leap forward in airplane design had hurled its airplane manufacturers into a deadly and unforeseen problem; however, this setback for British industry did not alter Trippe's plans for using jets. Many of Pan Am's engineers confirmed Trippe's decision by their rejection of conventional propeller planes and even the newer and faster turboprops, which represented a half-step between pure propeller and pure jet planes.[69] Props and turboprops alike were at the source of "an enormous percentage of mechanical breakdowns. No one wanted anything further to do with them."[70] If jets could increase passenger loads and simultaneously reduce maintenance costs, they would solve two problems for Pan Am at once.

Again, Trippe found ways to force American manufacturers to fulfill his airline's needs. While Lockheed was not interested in making jets at the time, both the Boeing and Douglas aircraft companies had plans for jet craft—plans that Trippe thought were inadequate. Trippe had learned about the new Pratt and Whitney J-75 jet engines, which were more powerful versions of the

J-57 engines that powered the air force's formidable B-52s; these engines were ideal to support large planes during transoceanic flights.[71] In a bid to force manufacturers to use the engine, Trippe actually bought 100 J-75s *before* any plane existed (even on blueprints) that would use such an engine. Trippe practically ordered Boeing's CEO William Allen to use them. "If you won't design a plane for the engines," Trippe threatened, "then I will find someone who will."[72] Allen had already invested a lot of money into another modern passenger plane and refused.

In the meantime, Donald Douglas was planning a new jet aircraft, the DC-8. Douglas was more accommodating to Trippe's demands and committed his company to making a version of the DC-8 that would use the J-75; these changes garnered an order from Trippe for 25 planes. Immediately afterward, Trippe ordered 20 Boeing 707s to be delivered in October 1958 even though the planes included the inferior J-57 engine. Both CEOs, thinking that they had scooped Pan Am's business, were surprised to learn from the pages of the *Wall Street Journal* on October 15, 1955 that Trippe had scooped them.[73]

With these orders in hand (financed primarily by a $60 million loan from the Metropolitan and Prudential Life Insurance Company where Trippe was a member of the investment committee), Trippe invited European colleagues in the IATA who had been such a hindrance to him to a meeting at his New York apartment on October 13. Serving his revenge cold, Trippe quietly made the rounds, mentioning here and there among his guests that he had just ordered 45 new jet aircraft. The news was staggering: the cost for both the Boeing and Douglas aircraft totaled $269 million. Now, with Trippe having set a new standard for technology in the airlines, foreign airline companies would have to play by Trippe's rules and collectively invest billions of dollars in jets or face certain demise competing with propeller-driven aircraft.[74] Trippe's bold tactics certainly had their effect—foreign aircraft carriers bought the new technology so quickly that at least 745 jets were in service all over the world by 1962.[75] Once again, we see evidence of a business leader fundamentally shaping the competitive landscape of an industry. Trippe's efforts to push the boundaries of aircraft technology resulted in quantum leaps in the distance that planes could traverse and the overall economics of passenger travel. To remain competitive, other airlines had no choice but to follow Pan Am's lead.

The beginning of the age of the jet also marked the end of the age of the subsidy for Pan Am. While in 1950 one-fourth of the company's operating revenues were mail payments, by 1956 the CAB cut Pan Am's postal subsidies completely. Much of Pan Am's European competition, on the other hand, still could count on very deep government pockets to sustain their operations. In the case of France, one historian has noted that between "1945 and 1965, Air France remained unprofitable." Government subsidies kept the airline going during this period. Air France was so protected by the state that it had the option "not [to] pay attention to commercial competition. . . . Air France commercial policy was based on its worldwide presence: 'in all skies' or 'the longest network in the world.' It was a public service more than a commercial firm."[76]

Trippe had bet Pan Am's future on its continued ability to grow and profit from economy- and tourist-class travel, a strategy that was dependent on being ahead of the competition with the best aircraft technology available. Despite the increased international competition, Pan Am thrived through the early 1960s. One historian commented that the "airlines seemed to be operating in Utopia."[77]

> Traffic was growing by about 15% a year, profits were rolling in, airline shares were favorites in the stock market and there appeared to be no end in sight to the era of plenty. The bonanza created one problem. Skies and airports were congested with planes—military planes, corporate and private planes, and scheduled airliners. At 33,000 feet above the North Atlantic, the airways were clogged with jet traffic. Forecasters sighted a 200-percent increase in the number of air passengers by 1980, and a proportionately larger volume of freight.[78]

Another Quantum Leap in Technology

As a final flourish to his reign over Pan Am, Trippe decided to repeat history a third time by prompting yet another quantum leap in airline technology. To stay ahead of the curve, Trippe believed that the time was right to invest in larger jets that could accommodate more passengers and fly at accelerated speeds. In early 1963, Trippe considered investing in the Concorde, a supersonic jet designed to fly at Mach 2.2 (1,450 miles an hour). Though Lindbergh counseled him against it, Trippe made his own assessment of the Concorde and decided to place an option on eight planes (later increased to fifteen). Though he was one of the first airline CEOs to order the new supersonic jet, it was not scheduled for delivery until the early 1970s, so Trippe sought additional options. He made the rounds to the major aircraft carriers—Lockheed, Douglas Aircraft, and Boeing—and in December 1965 signed an agreement with Allen of Boeing to develop what became the wide-body 747. The contract for the 747 included the following specifications: "gross weight of 550,000 pounds, capacity for 350 to 400 passengers, 5,100-mile range with full passenger payload, initial cruise altitude of 35,000 feet, flying at Mach .9, or just below the speed of sound."[79]

At a cost of $15 to $18 million each, Boeing believed that it needed to sell at least 50 to justify the massive investment in new plant and equipment. Trippe agreed to order a minimum of twenty-five 747s with the stipulation that Pan Am would receive delivery of their aircraft between September 1969 and May 1970, well ahead of competing airlines. Pan Am's order, which totaled approximately $500 million, was the "largest single undertaking ever carried out by a commercial company."[80] Under the conditions of the purchase agreement, Pan Am was required to make payments during the four-year period in which the planes were scheduled to be built.[81] At the time of the Boeing 747 purchase, Pan Am was already committed to orders for redesigned Boeing 707s and other aircraft. The combined total of aircraft orders was close to $1 billion.[82] To secure the funding for the new equipment, Trippe renegotiated loan agreements for $140 million, secured a line of credit for $100 million, and floated a $175 million issue of convertible debentures based on the booming Pan Am stock price.[83] The rest of the funding was predicated on additional earnings from the expected growth in passenger traffic in the airline industry.

Within two years of the signed contract for the Boeing 747, a number of unsettling events unfolded. Through a series of design changes (requested by Trippe and others at Pan Am), the 747 had ballooned to more than 710,000 pounds and the original engines were underpowered to accommodate the desired range for intercontinental flights. With the original engines and the higher weight, the 747 had the capacity to fly at a full payload from New York to Paris, not the much longer distances that were expected. In essence, "Pan Am was presented with an ideal plane for its competitors to operate on their high-density, short- to medium-range domestic routes. Instead of a great leap forward, the 747 represented for Pan Am an economic step backward; its performance level was inferior to that of the 707–321B [the latest modification of the Boeing 707]."[84]

Unlike their previous efforts to upstage the competition with access to advanced aircraft months or years before other airlines, the problems with the 747 essentially placed all airlines on a level playing field. More significant than the engine issue, the projections for increased international passenger travel through the remainder of the 1960s proved to be overly optimistic. Pan Am was set to take delivery of a number of jumbo jets at a time when many of the existing jets were flying half empty.

Trippe Steps Down

Trippe stepped down from the active management of Pan Am in 1968 at a time when the future of Pan Am still looked promising. The company's revenues surpassed the $1 billion mark for the first time and net income, though down from 1967, was a relatively strong $44.8 million.[85] At the time of Trippe's retirement, Pan Am had a route system of more than 80,000 miles and provided services to almost every airport in the world.[86]

Although the company was beginning to struggle with its massive debt load, Trippe believed that the gamble for the firm's future was still reasonable. Pan Am's troubles, however, were magnified in 1970 as it took delivery of the first 25 wide-body 747s (the company ordered 33 in total); at the time, the country had slipped into a recession precipitated, in part, by a crippling oil crisis. Trippe's biographers note:

> All the assumptions on which Trippe made his grandiose decision were turned upside down. Instead of a 17-percent growth in traffic, the rate of increase slipped to 4.6 percent, while Pan Am had 15.8 percent more seats to fill. The vaunted productivity gains of the 747 were predicated on expanding loads. Unanticipated rises in labor and maintenance expenses and the end of the era of cheap energy caused the direct operating costs of the superjets to soar rather than plummet.[87]

Through many changes in executive leadership as we will see in chapter 8, the company attempted to survive, but it was ultimately unable to cope with the even greater level of competition that resulted from the Airline Deregulation Act of 1978. For Trippe, the deregulation act was the epitome of his worst fears—massive competition, the exact opposite of a "chosen instrument" strategy. Trippe's legacy did not live on through Pan Am but incongruously, through the Boeing 747. Harold Evans notes:

> [The Boeing 747] came to the rescue of mass air travel at a time when the skies were dangerously congested; it effected a 3.5-to-1 increase in unit productivity while eliminating smoke and reducing noise pollution. It changed the economics of air travel and the nature of international tourism making both feasible for ordinary Americans. Trippe's act of faith in aeronautics was the greatest ever made by an airline in technology.[88]

Pan Am was unable to garner the benefits of the technology in time to salvage its own operations. Reflecting on Pan Am's missed opportunity, Richard Branson of Virgin Atlantic wrote: "Trippe had been a continuous innovator, but the sad irony is that he failed to reinvent his company for the leaner, far more competitive age he had done so much to shape: the age of travel for the Everyman."[89]

CHAPTER 5

C. R. Smith and American Airlines

As the U.S. government stepped in to formally regulate the airline industry in the late 1930s, a massive consolidation effort ensued that resulted in a streamlining of operations, a standardization of equipment and processes, and a relatively stable playing field. In addition, these efforts by the government to control the airline industry created the conditions that defined the dominant, domestic business model—one that was based on scale, scope, and reach. It was a model that heavily favored large corporations with deep pockets such as American Airlines. During the course of his 30-plus years as the president and chairman of American Airlines (1934–1968), Cyrus Rowlett Smith helped to create one of the largest airline carriers in the world.

Shortly after Smith became president of American Airlines in 1934, he took on another great responsibility: marriage to Elizabeth Manget. Active in the Dallas Junior League, Elizabeth soon found that her married life would become lonely and frustrating. While the Junior League exposed Elizabeth to many aspects of Dallas's civic life through the organization's commitment to volunteerism, "C. R.," as Smith liked to be called, had one overarching priority in his life: a devotion to American Airlines. Although C. R. certainly had great social skills, his single-minded devotion to his company meant that American Airlines' employees and customers received the bulk of his attention.[1]

One dramatic moment illustrates C. R. Smith's priorities. When Smith's wife went into labor in December 1939 with his only son, he made his way to Dallas to go to the hospital. On route from New York, he landed in Nashville where he was greeted by American Airlines employees who thought that he had decided to come to their Christmas party. Smith stayed for the party and then made his way to Dallas. One of the Nashville employees recounted: "he simply chose what to him was more important—he couldn't really help much at the hospital, but he figured he could do an awful lot of good at the party."[2] By 1941, Smith and his wife filed for divorce. His wife recalled: "I love the man, but I can't be married to an airline."[3] Smith remained single for the rest of his life.

Although this story might imply that the airline executive was terribly out of touch with life around him, Smith was extremely sensitive to the needs of his company and involved himself intimately in all its workings. As a steward of American, Smith chose to focus his energy and attention on his company—rather than balance the life of a CEO and a family man. For all Smith's success, he remained grounded. Smith typed his own memos, answered his own phone, and enforced a similar orientation among his managers, torpedoing any who took on "executive" airs.[4] Although blunt with executives, Smith was unassuming with customers and rank-and-file employees. He customarily introduced himself to every American ticket agent or baggage handler he met (and recalled details of their family history in the next conversation), not to mention to stunned passengers on flights where he wandered up and down the aisle, mingling and presenting himself as "president of this rodeo."[5] Eventually, Smith was flying 100,000 miles a year to keep abreast of the sprawling American network of routes.[6] The *Saturday Evening Post* described Smith's approach: "Often, on getting into one of his route towns, Smith invites the entire local staff, from manager down to washroom attendant, to a dinner at his hotel. Over the cigars and coffee, he holds a hair-down seminar on what is wrong and what could be done to improve it."[7]

This meticulous attention to customer service helped Smith to anticipate most of his customer's needs and to serve them through bold marketing, attention to safety, and shrewd applications of new technology. Smith recognized early on that customer service could be one of the few differentiators for American Airlines within the context of a regulated industry. Government regulation imposed restrictions on the routes that airlines could fly, the prices that they could charge, and the equipment they could use. As a skilled manager, Smith was able to work within these parameters to distinguish his airline in an otherwise static competitive landscape. By engaging employees and focusing on customer service, Smith was able to propel American into an industry leader.

Smith's Early Years

Smith's ferocious work ethic and attention to detail have their roots in his childhood when he and the rest of his family were forced to fend for themselves. Born in 1899 in Minerva, Texas, Cyrus Smith went to work around age nine as an office boy in Amarillo, Texas, after his father walked out on their family, in which Cyrus was the oldest of seven children.[8] Smith's mother Minnie took in boarders and taught school to make ends meet while all the children supplemented her earnings by working as soon as they were able.[9] The pooled resources, which went into a fund that Minnie called "The Smith Family Cooperative," provided the means that ultimately sufficed for all seven to attend at least some college, although Smith had to gain entrance via a special exemption since he had never finished grade school.[10]

Minnie also became an active participant in state politics, a venue wide open to volunteers, regardless of their means, and apt to bring one in contact with persons of influence. The Texas suffrage movement, in full swing during the 1910s, gained women the right to vote in state primaries by 1918, and Minnie's work brought her into the acquaintance of Pat Neff, governor of Texas in the early 1920s. Neff made a name for himself as a pioneer in appointing women, Minnie among them, to state boards and staff positions. He would also prove helpful later in getting her children started in careers.[11]

In addition to profiting from his mother's political connections, C. R. showed every sign of being well-equipped to manage his own career. Disgusted by a "Christmas bonus" consisting of a box of 10-cent cigars after he had spent most of the year working 15-hour days to take up the slack left by staff serving in World War I, he resigned from his job as a bookkeeper.[12] He had already made his own Texas political connections in addition to his mother's, and he shortly gained an appointment in a tax department under the secretary of state. When he did enroll at age 21 to study business at the University of Texas, he had a part-time job as an examiner for the Federal Reserve Bank also lined up. By his junior year he was class president; that same year he demonstrated his skill in the field of marketing, operating a side business that distinguished him as a pioneer in direct mail list brokerage under the name, C. R. Smith & Company.[13] Between his various business pursuits, Smith was earning $300/month, which was twice the income he made as an entry-level accountant with the Dallas firm of Peat, Marwick, Mitchell & Company, a position he chose to explore instead of finishing his studies at the University of Texas. Despite his youth, Smith's entrepreneurial experiences convinced the firm's partners that he was capable of managing a variety of complex clients. As a junior accountant, Smith was assigned to audit hotels, apartment houses, movie theaters, oil companies, and insurance providers.[14] Smith proved to be a very capable employee and was soon promoted to senior accountant for the firm's public utilities clients. It was in this capacity that Smith found his next opportunity.

After two years with the accounting firm, he left to work as an assistant treasurer for one of his clients, the Texas-Louisiana Power Company. Over the next few years, the firm's president, A. P. Barrett acquired a number of companies, including the Texas Air Transport Company that had a contract for delivering mail between Dallas-Ft. Worth, Brownsville, and Houston. Barrett bought a few more small airlines and consolidated them into the Southern Air Transport Company that carried both mail and passengers; he then asked Smith to oversee the finances of the newly consolidated firm.[15] Smith was not enamored by the opportunity, saying at the time: "I'm not interested in aviation."[16]

Despite his initial protests, it was not long before the challenge of the aviation industry and even flying itself (he trained as a pilot in his spare time) captured him. He made a deal with the general manager of Southern Air Transport: "You know something about flying. I know something about accounting. I'll teach you accounting if you teach me to fly."[17] Though Smith did learn to fly well, his business skills were far superior. Within two years, he had replaced the general manager (his flying instructor) as the head of Southern Air Transport, and, in that capacity, he helped to pioneer passenger air travel in Texas. Over the next year, Barrett acquired two additional airlines and merged them into a regional passenger and transport line that was bought by industry consolidator Aviation Corporation (AVCO) in 1929.[18]

Formation of AVCO and American Airways

AVCO originally entered the industry as a holding company with a desire to capitalize on the surge of investment and capital flowing to new aviation interests in the wake of Lindbergh's solo flight across the Atlantic Ocean.[19] At its incorporation in 1929, AVCO sold two million shares of stock at $17.50 per share, raising $35 million.[20] AVCO used these funds to acquire several companies in the expanding airline industry. One of these acquired companies was Southern Air Transport, the firm for which Smith was the general manager. The other two holding companies were Colonial Airways (Juan Trippe's former company), serving the northeastern section of the United States and Universal Aviation Corporation, serving the Midwestern states. Smith described the loose affiliation of airlines under AVCO: "The subsidiary airlines operated independently, and often competitively. The routes went from Coast to Coast, but not in a straight line, and connecting routes were a rarity. There was only one way to the West out of New York City for American Airlines' predecessors—and that was northbound through Albany and Buffalo to Cleveland."[21] After a year of massive acquisition spending, AVCO was faced with a $1.4 million loss and a vast and disparate array of disconnected airline operations. Through much haggling and negotiation, AVCO's 65 member board of directors voted to consolidate all airline service operations into one subsidiary. It was through this process that American Airways, Incorporated came into being on January 25, 1930.[22]

Three months after the formation of American Airways, the company received a tremendous boost from the postmaster general of the United States, Walter Folger Brown, who worked to consolidate the various entities that had secured airmail contracts from the Post Office Department at the infamous "spoils conference." Brown eliminated many small, regional carriers from the bidding process including Delta Air Lines (see chapter 3 on Woolman and Delta for more details) and awarded the bulk of mail contracts to three primary carriers who had the capability to traverse the continental United States—United, TWA, and American Airways.

Power Struggle at American Airways

When AVCO acquired Southern Air Transport, Smith went with the business, becoming a vice president in charge of the southern division and later head of nationwide operations for the newly christened American Airways. The company struggled to survive over the next three years and by March 1932 had gone through $38 million, which was $3 million more than its initial capitalization.[23] As the company tried to stem the flow of red ink, American Airway's first president, Frederic G. Coburn resigned and a battle ensued for corporate control between LaMotte T. Cohu of Air Investors, one of the original underwriting investors in AVCO, and Errett Lobban Cord who sold two small airlines (Century Air Lines in the Midwest and Century Pacific Lines on the West Coast) to the consolidator and was one of its principal shareholders.

Cohu initially won the battle for control replacing Coburn in the spring of 1932, but the war with Cord only intensified. During the next 9 months, Cord waged a proxy fight for control of the airline and eventually succeeded in owning 55 percent of the outstanding shares in the company.[24] After gaining the majority stake in the company, Cord became chairman of the board of directors, and his first act was to fire Cohu and install a new president, Lester D. Seymour.[25] Throughout the battle for corporate control, Smith backed Cohu and when his candidate was fired, Smith also suffered. He was demoted from head of nationwide operations and sent back to head the southern regional office.[26]

Despite his animosity toward Cord and the new president of American Airways, Smith stayed with the company. Over the next two years, Smith worked to enhance the airline's fleet. Marketing air travel was C. R.'s strong suit throughout his tenure at American, a talent he demonstrated in his first major decision while working under E. L. Cord: the introduction of Pullman-style sleeping compartments. It seemed that increased speed offered by airplane travel had not yet succeeded in drawing in the American public in great numbers to airplanes as an alternative to railroad travel. This was especially true during the early years of the Great Depression, when travelers had less money to take a chance on this new mode of transportation. C. R. was inspired to experiment with sleeper service after American's competitor, Eastern, bought the Curtiss-Wright Condor.

Most airline officials were disdainful of the early version of the plane. The president of Western Air Express declared: "The only way I'd buy one is if they built a tunnel through the goddamned Rockies!"[27] Slow, clumsy, loud, and inefficient, the long Curtiss-Wright biplane was anything but elegant; but it *was* roomy. C. R. saw that as a competitive advantage, and he championed acquiring an improved version of the aircraft that would feature more insulation, less vibration, and engines that were much more powerful and responsive than Eastern's Condors.[28]

Bringing Marketing into the Mix

Unlike other planes, the Condor could easily accommodate 14 sleeping berths. These ample accommodations, unique in the airline business, were also described in American's advertising as a quantum leap from transportation via rail. Instead of "the noise, dirt, and confusion generally associated with ground transportation," travelers were encouraged to "fly in bed" to enjoy the heavenly influence of airplanes: "As you sleep, your lungs gratefully drink in cool, fresh uncontaminated air.... When you wake in the morning, you will say that never (no, not even in your own home) have you slept so restfully, so peacefully, so undisturbed." Although the advertisement was guilty of overstatement (including listing a top speed of 190 mph, 50 miles above the Condor's realistic speed), the public responded with enthusiasm.[29] Stewardesses, first introduced by United three years earlier in 1930, enhanced the comfort of American Airways' passengers who seemed to agree with American's claim that the Condor "revises all previous conceptions of an overnight journey."[30] The introduction of services like the sleeper berths is the kind of incremental innovation or differentiation that enables growth and expansion during the managerial phase of a business.

American's Condor sleeper service was first introduced on the Fort Worth to Los Angeles route of the firm's transcontinental route. Neither United nor TWA had anything similar. With the new Condor service, the performance of the southern division of American Airways far surpassed all other divisions and significantly enhanced Smith's reputation and profile in the company. Although Cord may have been displeased with Smith's lack of initial support, he was impressed with the newfound opportunities for enhanced revenue.

Although Smith would later be known as a staunch advocate for increasing the safety standards for airplanes across the airline industry, at this early stage in his career he seemed somewhat unconcerned about the Condor's shortcomings. As much as passengers enjoyed the ride, the plane was unusually light for its size, causing it to buck dangerously during bad weather. Pilots vociferously complained about the tendency of the plane's wings to gather ice on its northern journeys. And while the plane could be dragged down by ice outside, inside there lurked a high risk of fire.[31] Even though AVCO's chief safety officer Bill Littlewood despised the Condors, Smith kept with them.[32]

Air Mail Act of 1934

At the same time that American Airways was creating a new revenue stream through the sleeper service, it faced a major challenge to its airmail contracts. As discussed earlier (see Chapter 3), President Franklin D. Roosevelt, under pressure from an investigation into the awarding process of airmail contracts at the spoils conference of 1930, issued an executive order to cancel all existing contracts in February 1934. As a result, airlines were forced to furlough hundreds of workers and most schedules were pared down to passenger-only service (see sidebar).[33]

Braniff Airways: A New Regional Economy

When Postmaster Walter Folger Brown took the raw clay of the early airline industry into his hands during the "spoils conference" of 1930, he shaped it into a transcontinental route system configured in an east-west direction. Other major air routes favored the population and economic centers of the East Coast and West Coast. In between, most small airlines existed as feeders for the three large domestic airlines—TWA, United, and American.

In contrast to Brown's grand vision, Thomas E. Braniff began a small airline in Oklahoma that would grow largely independently of the major airlines in a north-south direction. Unanticipated by Brown, Braniff Airways emerged in 1930 with air passenger service connecting small towns—underserved by the larger airlines who focused on their east-west market share—whose economic fortunes were being transformed by the emergence of a new valuable commodity: oil.

Now, suddenly, poor towns in Kansas, Oklahoma, and Texas whose shaky economic histories had left them with an infrastructure of bad roads and slow railway service to connect them to the outside world were joined by the burgeoning petroleum industry. There was no time to make roads and rail connections that could serve the needs of that industry in the early 1930s; luckily, with Braniff Airways providing air passenger service to the center of the country, there would be no need for them.

In contrast to the family-oriented marketing of C. R. Smith's American Airlines, which emphasized comfort and safety, Braniff Airways was known early-on for speed, which was exactly what people in the early oil industry needed to close fast-moving deals and to look for new places to extract the black gold. Braniff advertising encouraged passengers to "take the B-Line," "the world's fastest airline." Slowly but surely, Braniff grew until, by the beginning of World War II, it could describe itself as the airline that offered service from the "Great Lakes to the Gulf."

Although Braniff certainly could have attempted to emerge as a competitor with some of the bigger airlines in the 1930s, he kept a relatively low profile before World War II. Anticipating unfriendly attempts by larger airlines to acquire his company in the immediate future, Braniff took the lead in founding the Independent Scheduled Air Transport Operators' Association that joined many small airlines working in the South and Southwest. A center for resources and information, the association helped to stop potential in-fighting among these small companies and also provided them with important legal representation in Washington, D.C.

Braniff's long-time presence in the country's oil region eventually prepared the way for his airline to enjoy significant growth along with the petroleum industry during the postwar industrial expansion of the economy.

The Air Mail Act of 1934 spelled out the provisions of the new award process as follows: (1) all aircraft manufacturing companies must divest themselves from their airline subsidiaries; (2) no airline executive who attended the spoils conference would be allowed to hold office in a carrier possessing a mail contract; and (3) no airlines that had participated in the conference would be permitted to bid for any mail contract.[34] In accordance with the provisions of the Air Mail Act of 1934, American Airways divested itself of the few aircraft manufacturing units that Cord had brought with him or purchased as head of the company.[35] The second condition of the Air Mail Act caused no real concern for American; only one minor executive that was present at the spoils conference of 1930 was still around, and he was forced to resign. To comply with the third major provision of the Air Mail Act, American Airways simply changed its name to American Airlines. After mail contracts were reopened to private airlines a few months later, American, now reconstituted as American Airlines, won its fair share of the rebid contracts, and some suggested perhaps more than its fair share. There were 43 qualified bids for 21 airmail routes; American bid on eight of them and won most of its former routes.[36]

American's success in winning back most of its old routes was due in no small part to one of C. R.'s abiding strengths as a CEO: good relations with government officials. Like his counterpart Juan Trippe, Smith was personally involved in cultivating connections within the government. Through the rebid process, American actually secured routes that created a more efficient transcontinental route, which caused *Fortune* magazine to report that "American was the only transcontinental airline to come out of the post-cancellation bidding with a better system than before."[37] Though its route system was better, American settled for a much lower rate for carrying the mail. American was paid an average 42.6 cents per mail mile in January 1934; by July, the rate was set at an average of 25.3 cents per mail mile.[38]

As the new bidding process unfolded, Cord became convinced that American Airlines needed a new leader—one who could expand the company's passenger service in light of the reduced revenue potential for airmail contracts. On October 26, 1934, the board of directors formally elected Smith to the presidency of American Airlines.

Smith Assumes Presidency of American Airlines

At the time that Smith assumed the presidency of American Airlines, it was the largest airline in the United States in terms of routes and equipment. American also had the unfortunate distinction of carrying the largest operating loss, $2.3 million in 1934, and having the worst reliability record.[39] The state of their equipment was no better. Years later, Smith recalled: "At the time of the reorganization in 1934, the company had a fleet of Condors, and Fords, and Lockheeds and perhaps a half-dozen other types. The disparities among models made them basically uneconomical to operate. What was needed was one attractive type capable of carrying a sizable load a good distance, quickly, and at low cost."[40] Despite the relative success of the Condor for American's bottom line and despite the valiant efforts of people like Harry Guggenheim, the public in general was still not completely convinced that flying could be a normal, everyday activity. As American Airlines biographer, Robert J. Serling notes, in 1934, "fewer than 500,000 Americans had ever been in an airplane—less than one half of one percent of the population....99.5 percent stayed away from flying mostly because of fear, and while cost played some role, it was secondary... Most life insurance policies were automatically suspended during the time a policyholder was in any airplane, commercial or otherwise."[41]

To build consumer confidence in flying, Smith believed that American Airlines had to apply the focus on safety to a new and more efficient fleet of airplanes. Though American had built a viable transcontinental service with the use of the Condor, its limitations were great. Even worse, other airlines had far surpassed its performance by the time Smith became president. TWA had introduced the DC-2 aircraft to its fleet in July 1934 and boasted that it could offer coast-to-coast service in 15 hours, much faster than the Condor. Despite the size of its fleet, American's equipment was considered obsolete and pilots often joked that the "airline was called American because of its democratic equipment—you can fly anywhere in anything."[42]

The DC-3 Decision

Though Smith ordered a number of DC-2s to supplement American's base of equipment, the order was placed six months after TWA. Hoping to leapfrog TWA and other carriers, Smith pushed Douglas Aircraft to develop a new plane (ultimately the DC-3) that would offer 50 percent greater passenger capacity than its previous model (DC-2) with better fuel performance. Smith's requirements for the new plane also included his vision of an airline that would be free of its dependence on unpredictable government subsidies. Smith planned for the new plane to carry 21 passengers in the daytime and 14 at night. He believed that the plane that he was requesting from Douglas Aircraft would be 85 percent DC-2 and only 15 percent new. In actuality, the ratio was reversed.[43] Douglas Aircraft initially balked at the request; it was having difficulty managing the demand for the DC-2s and saw no need to experiment with and invest in a radically new aircraft.

Smith's perseverance in the face of Douglas's resistance was derived from the belief he had in his own vision of the future needs of the airline industry and his ability to use his connections with government officials to help his company's finances. Douglas was finally persuaded to produce the DC-3 after Smith made a commitment to buy the new model in volume (20 immediately and an option for 20 more). The 40 planes would allow American to standardize its operations, a highly speculative move for a plane without an established track record of safe operation. The tentative price for the new DC-3 was set at $100,000 per aircraft for a total order of approximately $4 million.[44] This was quite a sum, especially since American Airlines did not have a lot of money. As a result, Smith said, "We bought the plane from Douglas without ever signing an actual contract. We did not want to sign one because it was customary to put up a cash deposit when you signed, and we didn't have enough cash to make the usual required deposit."[45]

Immediately afterward, Smith found a good way to compensate for American's tight cash flow. He flew to Washington, D.C. to visit Jesse H. Jones, a fellow Texan of Smith's acquaintance. Jones was the head of the Reconstruction Finance Corporation (RFC), a government lending agency founded in the last days of the Hoover administration and empowered with great discretion to extend credit to large banks and businesses. Getting to the point as only two Texans could, Smith started the conversation:

> "Jesse, I've read that the RFC was set up to ward off financial disaster."
> "That's right," Jones allowed.
> "Well, American Airlines is a disaster if you don't make us a loan."

Persuaded by Smith's reasoning, Jones quickly approved a $4.5 million loan for American Airlines.[46]

DC-3 service began in 1936 and one aviation historian called it "the most significant date in the first twenty-five years of air transport history."[47] The aircraft held 21 passengers, cruised at a speed of 195 miles per hour, and flew 1,000 miles without refueling. The DC-3 quickly surpassed the current state-of-the-art aircraft that was manufactured by Boeing Corporation, which could hold 10 passengers and cruise at 165 miles per hour.[48] The DC-3 could also be built quicker and cheaper than any other aircraft enabling American to efficiently expand its operations. By standardizing its fleet, American was able to vastly improve its reliability and operational efficiency for service and maintenance. The airline no longer had to maintain spare parts inventory for multiple aircraft models, and it was able to vastly reduce the training of mechanics.[49] Smith recalled: "The DC-3 was, for its day, the perfect transport. It struck the balance that airline engineers search for in speed, in gross weight, in power and payload space and wing area. It permitted economies that had been beyond the company before."[50] Serling recounts the impact of the DC-3 on the airline landscape: "The impact of the DC-3 can best be measured not by the number of curious spectators at static exhibits but by the number of people who began flying them. In 1936 alone, the U.S. airlines for the first time in their history carried more than a million passengers—doubling the 1934 total—and the traffic curve was to mount steadily from then on."[51] Passenger revenue for American Airlines increased from $1.9 million in 1933 to $5.6 million in 1936. In contrast, airmail revenue for American Airlines decreased from $4.7 million in 1933 to $2.0 million in 1936.[52] The increase in passenger revenue experienced by American Airlines was part of an overall pattern in the industry (see figure 5.1).[53]

Smith's gamble on the DC-3 purchase helped to catapult American Airlines to the forefront of airline technology in the United States. By the end of 1936, American had secured the first 20 DC-3s that it had ordered. In contrast,

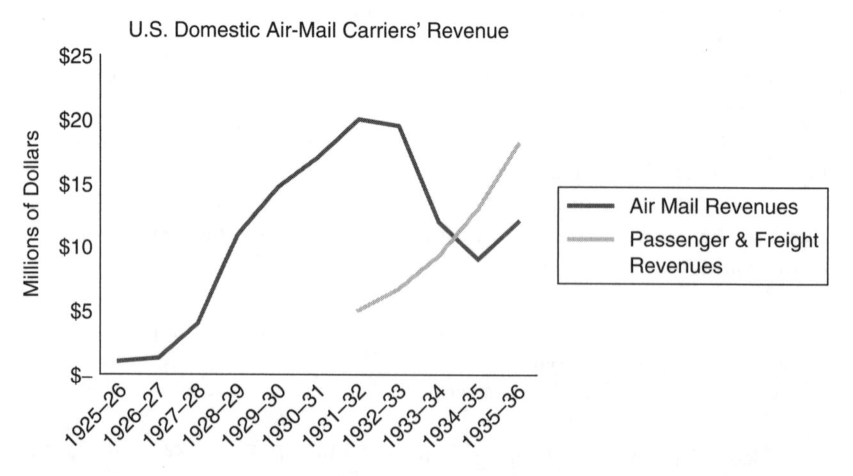

Figure 5.1 Passenger revenue outpaces airmail revenue by 1934

United had only ten, and TWA was still waiting for its initial order of eight to be delivered.[54]

To showcase the DC-3 to the American public, Smith's public-relations team staged a major publicity event. In its inaugural flight, the DC-3 "took off from Chicago with a fuel load of 822 U.S. gallons, flew to Newark without stopping to land, and went right back to Chicago's Midway—a nonstop round trip completed in eight hours and seven minutes."[55] The speed with which the aircraft completed this roundtrip was considered unprecedented and garnered a tremendous amount of attention for American though little notice was paid to the fact that there were only three passengers on the flight and no mail (a full load would have resulted in far different results). Publicity became a major component of American's strategy as it attempted to significantly increase the number of airline passengers.

The diminishing revenues associated with airmail service made it all the more important to attract passengers. Attracting passenger traffic depended on (1) obtaining routes that were part of a comprehensive network; (2) increasing the speed of aircraft to reduce travel times; (3) enhancing passenger comfort; and (4) improving safety to reduce consumer's perception of risk. Even after more than a decade of passenger flights, consumer fears became one of the most intractable obstacles to building the base of passengers.

American Airlines' DC-3 airplane waits for take-off on the runway. (Source: Bettmann/CORBIS).

Combating Consumers' Fears of Flying

Although the success of the DC-3 is now part of the lore of modern technology, its triumph beginning with American Airline's large purchase in 1936 could have been squelched—not because of any great mechanical shortcomings, but because of an unfortunate rise in highly publicized airplane accidents. Notwithstanding the merits of the DC-3, fear still entered the mind of most travelers contemplating transportation by air. In 1936, airlines were experiencing an accident rate of approximately one every six weeks; this culminated in a spike of five accidents occurring in December 1936 and January 1937. Although this airplane accident rate was not extraordinary compared to other modes of transportation, the press amplified the crisis with spine-chilling headlines: "All on Plane Dead" and "Bodies of 12 on Giant Air Liner Found in Deep Ravine" were typical fare.[56] The results were immediate. During that winter of 1937, thousands of passengers cancelled their airplane reservations (see figure 5.2).

The problem of safety had now become paramount to the industry. Most airline presidents, like Pat Patterson of United, tackled the problem principally by internal improvements in their company's maintenance and safety procedures.[57] In contrast, Smith took the much bolder step of using his marketing skills to turn a public-relations disaster into an advertising coup. Taking a page from the playbook of Harry Guggenheim, Smith began an advertising campaign in 1937 that focused on safety, a subject most airline executives believed was best discussed behind closed corporate doors. Instead of the standard advertising, which featured an airline's aircraft and a timetable of flights, Smith took the unprecedented step of addressing the taboo subject in a bold headline: "Afraid to Fly?" Immediately underneath, C. R. Smith's small portrait stares confidently at the reader, simultaneously reassuring the traveling public while putting his own reputation on the line. Although he is speaking as the head of American Airlines, Smith was in fact addressing the entire traveling

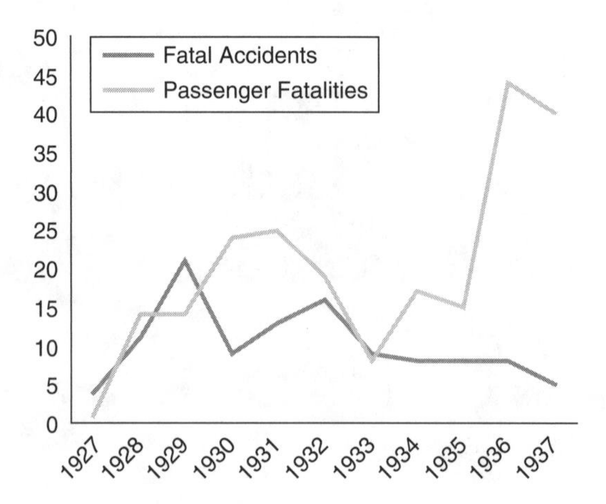

Figure 5.2 U.S. air transportation accidents: 1927–1937 (Air Transport Association of America).

public, no matter what their favorite airline was. With this bold advertising campaign, Smith became the unofficial leader of all the airlines, a move he felt he had to make to save an industry in crisis.[58]

Going one step further than Guggenheim, Smith eschewed the voice of the expert who proclaimed the results of his scientific investigations to the public. Instead, he took on the persona that he used with all his customers—confident, yet without airs, Smith sounded almost paternal. "The fact is," Smith confides in his customer, "there *are* risks involved in *all* kinds of travel" and statistics show that air travel is no more dangerous than any other. Then, leaving all pretense of expertise behind, Smith's writing took on a quiet, friendly tone, speaking man-to-man to the American public:

> Why quote statistics? They are not always conclusive. They are often only controversial. I could show you figures to prove that you would have to fly around the world 425 times—or maybe approximately 14,165 flights back and forth between New York and Chicago—before you would be liable to meet with an accident. Do these statistics overcome your fear of flying? I think not. There is only one way to overcome that fear—and that is, to fly.[59]

In using the first person, Smith shrewdly played on the cultural preferences of most Americans who often admired leaders who did not seem too different from themselves. The advertisement provided a significant boost for American, but it also was used by competing airlines to extol the virtues of flying. Many of these competing airlines simply reprinted the advertisement and distributed it to major businesses as a way of increasing overall travel.[60]

Smith even made safety the cornerstone of the company's employee magazine, which included the following axiom under its masthead: "Aviation is not unsafe, but like the sea it is terribly unforgiving of any carelessness or neglect."[61] American's focus on a safe and efficient fleet helped to generate $6.6 million in passenger revenue in 1937, which was twice the sum generated from airmail contracts.[62]

In launching this innovative advertising campaign, Smith saved his $4.5 million investment in the DC-3. But the quality of the new Douglas plane had made that $4.5 million a pretty safe bet. The DC-3, Smith noted, permitted airlines to do something that they had never been able to do before: "The DC-3 freed the airlines from complete dependency upon government mail pay. It was the first airplane that could make money by just handling passengers. With previous aircraft, if you multiplied the numbers of seats by the fares being charged, you couldn't break-even—not even with a 100 percent full load. Economically, the DC-3 let us expand and develop new routes where there was no mail pay."[63] To build upon the DC-3, Smith added further innovations in airline marketing that promised to encourage more and more people to make the airplane their preferred mode of transportation.

Smith's focus on passenger travel was further enhanced through the creation of discount fares and the first VIP passenger organization, The Admirals Club. American established the Air Travel Card system in the mid-1930s, which enabled travelers to receive a 15 percent discount on air travel in exchange for an annual deposit of $425. By the end of the decade, all major airlines had a similar discount card program to attract more frequent travelers.[64] The idea for the Admirals Club, a private room in major airports for VIP passengers to congregate and relax between flights, emerged after Smith was named an Honorary Texas Ranger. Reflecting on his award, Smith commented to his director of public relations: "You know, these guys in Texas and the ones in Kentucky, with their honorary colonels, they get a lot of mileage out of those little pieces of paper. Why can't we have something like that?"[65] What began as an honorary designation for key customers became an integral and exclusive club, one sought after by key customers. It became a new standard for all airlines and a cornerstone of future frequent flyer loyalty programs.

Powered by the DC-3s and Smith's aggressive marketing tactics to build awareness and popularity of air travel, American was the only domestic transcontinental airline to avoid red ink in 1938, posting an after-tax profit of $213,262 in a year when United lost nearly a million dollars and TWA more than $750,000.[66] The same year, the remaining $1.4 million of American's $4.5 million RFC loan was paid down below $100,000.[67] Profits climbed to almost $2.5 million by 1941 and the employee base grew to more than 4,000, up from 2,795 in 1939.[68] During 1941, American was also recognized with a safety award for flying a billion passenger miles over the course of five years without a single fatality.[69] At the time, American, as one of twenty operating airlines, was flying almost one-third of all passengers in the United States using sixty-four DC-3s and fifteen Douglas Sleeper Transports (modified DC-3s).[70] American reigned as the largest airline carrier in the United States for two decades.

Smith at Air Transport Command during World War II

Smith's accomplishments in overseeing the transformation of American Airlines were recognized by the U.S. government, and, in early 1942, he was asked to take a military post heading the domestic Air Transport Command (ATC) during World War II. The ATC was responsible for coordinating the use of commercial and military aircraft to transport soldiers and equipment throughout the globe, and all airlines were required to support this massive mobilization effort. For its part, almost half of American's fleet of equipment and personnel were under contract with the War Department between 1942 and 1945.[71]

Unlike Juan Trippe, who strongly resisted the idea of formally joining the armed forces during the war, Smith went to the military very willingly, a gesture that certainly did him no harm in the halls of government. Though Smith made an effort to not favor American Airlines or individuals connected with his former company in awarding air travel contracts, by the end of the war, American had become the second largest international air carrier in the world (topped only by Pan American) with more than 11,000 personnel and close to 100 DC-3s.[72] During the war years, American generated $16.5 million in profits that was entirely derived from commercial operations. All ATC contracts were performed at cost.[73] Smith was named a major general in recognition of his efficient oversight and management of the ATC.

American Airlines after World War II

Shortly before Smith returned to the helm of American Airlines in the summer of 1945, the company had acquired American Export Airlines (AEA), a small international air service carrier that had been established in the late 1930s to expand the transportation offerings of the American Export Line shipping company. The path to the acquisition was paved by the Civil Aeronautics Board (CAB).

As World War II drew to a close, the CAB sought to increase the level of competition for international airmail and passenger travel. The CAB not only approved American's acquisition of AEA, but it also authorized it to fly routes to Europe over the North Atlantic corridor.[74] At the same time, the CAB also approved TWA's request for European routes. The international routes approved for TWA and AEA did not initially come with corresponding airmail contracts; those contracts remained with Pan American. Hoping to expand its international network, Pan American fought unsuccessfully to acquire AEA. Ironically, it was Pan American who forced American Export Lines to sell AEA by successfully claiming in its court action that shipping companies were barred from owning other forms of transportation.[75]

When Smith returned to American, AEA's name was changed to American Overseas Airways (AOA) and it became the foundation for the transatlantic division of American Airlines.[76] In addition to integrating AOA into the American family, Smith's immediate postwar priorities consisted of equipment upgrades and route expansion. He believed that there would be a major boom in passenger air travel in the late 1940s and 1950s.

To support this expected increase in air travel, the CAB was inundated with new route applications. The domestic "Big Four" airlines—American, United, TWA, and Eastern—continued to receive the bulk of new routes, but many small, regional carriers including Delta, Continental, and Braniff were able to secure additional routes to expand their coverage base.[77] Leveraging his lobbying skills, Smith spent a considerable amount of time in Washington, D.C. during these years to ensure that American received its fair share of new route awards.[78] In 1945, American had a fleet of 86 planes, which covered 9,457 miles. Smith hoped to double the airline's route mileage through a series of petitions to the CAB.[79] To accommodate the anticipated increase in air travel, Smith once again dramatically expanded American's fleet of aircraft.

A New Fleet

At the end of 1946, Smith signed an $18 million contract for 100 Convair, CV-240s, which by accommodating 40 passengers almost doubled the capacity of the DC-3.[80] At the time, this transaction was the largest single aircraft order in commercial aviation history, but it was only one part of Smith's expansion plan. He had already committed the airline to purchase 50 DC-4s (a slightly modified DC-3), 50 DC-6s, and 20 Republic Rainbows for a total of 220 new aircraft at a cost of $90 million.[81] With this unprecedented overhaul of equipment, Smith hoped to retain American's position at the forefront of aviation technology and production. He also sought to derive maintenance and service efficiencies with a large standardized fleet, similar to his approach with the DC-3. To pay for the new equipment, Smith was willing to gamble on the company's future. He heavily leveraged the airline and sold $40 million of preferred stock and $40 million in debentures.[82] Reflecting on the investment, Smith asserted: "The investment seemed mandatory. The company had learned long ago that as the business grew so did its capital requirements. The DC-3, once the ideal airplane, was no longer economical to operate. Costs were going up, but fares were remaining fairly stable. What was needed were bigger and faster planes capable of carrying more people farther in less time."[83]

Smith's plans for growing his fleet were soon modified to respond to unanticipated opportunities and problems. Production problems plagued the construction of the Rainbows, and Smith eventually cancelled the order. The order for the CV-240s was reduced from 100 to 75 as Convair struggled to profitability build and deliver the aircraft. Smith had negotiated a unit price of $195,000 for the CV-240, and Convair was incapable of producing the aircraft at that price and hoped to cancel the entire order. Smith retorted, "We need the 240, so I'll make a counteroffer. We'll cut our order from one hundred to seventy-five and you can sell those twenty-five we're giving up for a profit." Convair accepted the offer.[84]

The postwar surge in passenger travel, however, did not occur at the rate that many airline executives, including Smith, expected that it would. Though airlines had made many strides in demonstrating the safety of air travel during the massive war mobilization effort, a series of high profile crashes and technical difficulties brought forth a renewed sense of fear and concern in the public. The air-conditioned DC-6, Douglas Aircraft's largest and fastest airplane at the time, suffered a number of embarrassing and fatal accidents that forced the CAB to ground the plane until the cause of the accidents was determined (a problem with the fuel transfer process).[85] TWA, which had relied on the Constellation aircraft designed by Lockheed instead of the DC-6, also experienced a high profile midair fire. As a result, "Manufacturers, the government, and the airlines alike shared the collective black eye inflicted by the Constellation and DC-6 grounding—a black eye whose shading turned even darker in the summer of 1948 when the new Martin 202 was grounded for major structural deficiencies . . . Airline payrolls became grossly top-heavy as anticipated traffic growth not only failed to materialize but dropped far below predicted levels."[86] This kind of optimism about demand is what buoys managers, but it often materializes in fits and starts, rather than a steady increase. The capacity to withstand this cyclicality, which requires betting on the underlying secular demand, is one of the hallmarks of good managers.

The heavy investment in new equipment combined with fewer-than-expected passengers resulted in some difficult financial years for American and other airlines. Though operating revenues increased from $47.4 million to $89.3 million between 1945 and 1948, American suffered losses of $0.4 million in 1946, almost $3.0 million in 1947, and $2.9 million in 1948.[87] By the end of 1947, American was forced to eliminate 15 percent of its employees. Despite the difficult situation, Smith remained "grimly optimistic."[88] Reflecting on this troubled time, Smith recalled: "General over-expansion of routes had brought about a very difficult competitive situation. Though the volume of traffic between cities was increasing, it was divided among so many carriers, that by 1948, it was difficult to maintain adequate load factors and frequency of service."[89]

Renewed Customer Focus

As he had done before, Smith focused on safety and public awareness. When the fuel transfer problem had been remedied in the DC-6, it enjoyed an enviable track record and performed to its early expectations.[90] In 1948, Smith introduced the family fare plan designed to encourage more air travel through reduced fares during nonpeak travel periods. Smith reasoned that a reduced fare would result in more travelers and thereby offset any potential decrease in revenue. In 1945, Smith had articulated his vision for the airlines in a *Saturday Evening Post* article, before problems with the DC-6 had become apparent:

> First, we need an air-line fleet so big that it constitutes an adequate reserve for national air power. Thousands of air-liners, not hundreds. Second, fares must be cut down to the pocketbook level of the average citizen. Volume business will result... The air lines are used to selling tickets to movie stars and big-business executives. We should be selling seats to the millions who have to pay for their own tickets out of middle-class incomes—the housewives, small businessmen, farmers, and mechanics.[91]

To better understand the opportunities for reducing fares, American conducted a load survey analysis for six months. Traffic was extremely high on Friday and Sunday. It dropped much lower on Monday and dipped even more on Tuesday through Thursday. And the lowest load factors occurred between Saturday noon and Sunday noon.[92] American offered 50 percent discounts for family members traveling with a full fare passenger during low peak periods. The response was an instant success. United had conducted a similar survey and, instead of reducing fares, it initially opted to reduce schedules. After seeing American's success, United and other carriers adopted a similar program.[93] By the early 1950s, Smith's investments in new equipment and marketing programs began to bear fruit. The company posted revenues of $118.7 million in 1950 rising to $428.5 million by 1960.[94]

As American Airlines' fortunes were rebounding, it had to contend with a whole new class of competitors called "non skeds." These were essentially charter airlines whose irregular schedules and service offerings (usually to one or more resort locations) enabled them to circumvent the bureaucratic oversight of the CAB. The charter airlines typically offered much lower fares than certified airlines and attracted a whole new clientele—just the type of traveler that Smith sought. At the time, all air travel on major carriers was offered at one level, first class, and its price was often out of reach for most middle-class travelers. The major carriers tried to regulate the charter operations by appealing to the CAB, but they were largely unsuccessful. Smith joined the fight, but he also recognized the potential opportunity for a "coach class" fare and began to develop a new offering for American Airlines.[95] In 1951, approximately 7 percent of American's passengers purchased coach fares. Three years later, coach passengers accounted for 25 percent of American's passengers.[96] To accommodate the coach service, American converted many of its DC-6 aircraft to carry up to 80 passengers.

This process of expanding the customer base is typical as industries mature. Initially, new businesses often serve a small, wealthy constituency at the top of the customer pyramid. As the industry grows, businesses expand the core offerings to a larger pool of constituents that often requires a dramatic reduction in prices. In essence, they move down the demand pyramid where the customer base is significantly larger (see figure 5.3). Both Trippe and Smith recognized the opportunities for expanding the airline industry by reaching out to a larger potential pool of customers, which, to be successful, required a reduction in prices.

As the airline expanded, Smith invested heavily in technology to improve the logistical tracking of flights for both American employees and passengers. In 1952, the company introduced the Magnetronic Reservisor to monitor the seat availability on its flights.[97] This system was modified and revised at the end of the decade with the assistance of IBM. After studying American's reservation problem for three years, IBM decided to commit itself to the project. The successor reservation tracking system was called SABRE (Semi-Automated Business Reservations Environment), and, at the time of its debut in 1961, it was the largest electronic data processing system used in business.[98] While the Reservisor

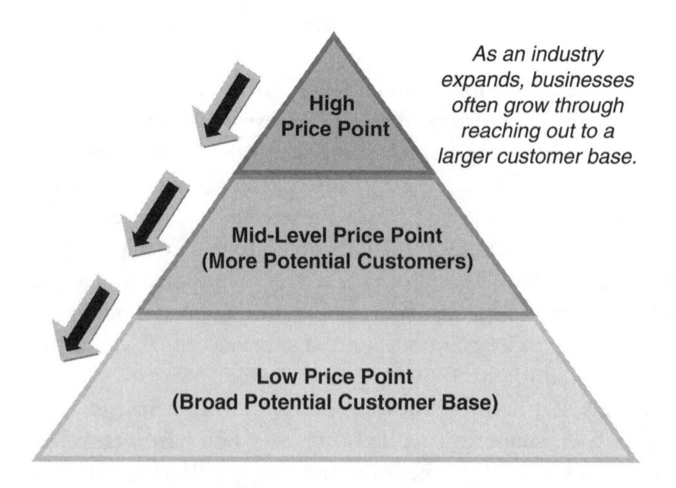

Figure 5.3 Customer demand pyramid

processed a reservation in 45 minutes, SABRE was able to do so in less than 3 seconds.[99] This was an amazing example of an internal business innovation that dramatically improved the operating performance of the entire industry.

The use of technology to create a more efficient business is typical of great managers. Building upon the entrepreneur's vision, the manager creates the infrastructure of an organization and the processes that make it run effectively and efficiently. Smith implemented a marketing campaign, initiated the development of sophisticated technological systems, and instituted an organizational structure to manage the complexity of the airline. Through these managerial efforts, American Airlines was able to thrive and grow as the airline industry matured. In essence, the manager's role is to regularize the business model so that it can effectively be reproduced on a much larger scale, and Smith was brilliant at this process.

International Travel: Smith's Missed Opportunity

In an effort to streamline American's operations, Smith decided to sell American Overseas Airlines to Pan American only three years after it had been acquired. Though from a customer's perspective, AOA's service to Europe was considered far superior to TWA's and Pan American's, the airline struggled financially. The airline was heavily dependent on the domestic division for marketing support, financial advice, and managerial oversight. Smith grumbled: "Every time I look around for one of my key officers to get something done around here, I find he's off somewhere in Europe working on some AOA problem. Management is spending 90% of its time on an operation that's producing only 10% of our revenues."[100]

Smith decided to sell at a time when American had suffered three straight years of significant financial losses. The CAB had eventually awarded AOA some overseas airmail subsidies, but they were not enough to compensate for the airline's other costs. Smith's decision to sell AOA was also influenced by the growth of national carriers in Europe. Concerned that the sale to Pan American would not be in the public's best interest, the CAB initially rejected it.

Since the awarding of international routes required presidential approval, the CAB's rejection was sent to President Harry Truman who initially endorsed it but later reversed his position and approved the sale.[101] The reversal was met with a considerable protest from TWA who sought to fight the decision by appealing to the Supreme Court. Smith's decision to maintain closer ties to Washington after the war undoubtedly helped to change Truman's decision. Smith's chief lobbyist helped to broker an agreement whereby Pan American acquired AOA and TWA obtained approval to operate new routes to Europe (specifically to London and Frankfurt).[102] After two years of wrangling, Smith finally sold AOA to Pan Am. Reflecting on the decision, Smith said, "on the basis of the situation at the time, we did right, I believe. On the basis of the situation many years later, the decision is debatable."[103]

If one looks at the opportunity costs of leaving the international travel arena in 1950, Smiths sale of AOA was more than "debatable;" it was, in fact, one of the major mistakes of his leadership of American Airlines. Although hindsight is often 20/20, Smith's impatience with the bottom line in the late 1940s made him unusually insensitive to signs concerning the growing potential of international tourism. From the perspective of the late 1940s, encouraging Americans to fly abroad was not an outlandish goal; Juan Trippe had already spent more than a decade doing just that.

Moreover, following the war, an increasing number of Americans possessed more disposable income, more leisure time, and more education than Americans had just a few years before. These changes provided the necessary foundation for international tourism. Reacting to these changes, domestic airlines were coming up with innovative domestic package vacations, which "offered one low all-expense rate for airfare, hotel, and ground transportation. In 1948, Delta introduced package vacations to Miami and United touted package tours to Hawaii." As early as 1947, *Better Homes and Gardens* informed its wide, middle-class audience that these package tours were going to "change many established travel habits. You'll find yourself planning in terms of where you'll really want to go, regardless of distance."[104] As we have seen, Smith had long been an advocate of creating innovative fare structures to increase passenger traffic on domestic routes; why was he not able to see the potential of similar tactics for international flights?

In contrast to Juan Trippe, who had long believed that international travel could be very lucrative, Smith may have lacked Trippe's entrepreneurial tolerance of risk that fueled Pan Am's chaotic early history. In addition, Smith may not have been as familiar with foreign countries as some of his contemporaries, such as C. E. Woolman and Harry Guggenheim who both had traveled and worked abroad as young men. And, unlike Trippe, Smith took the helm of American Airlines after it had already established a transcontinental presence; Smith may not have understood (at least not as well as Trippe did) the value of being among the first to exploit a new market. As a result, American missed out on the growing air passenger traffic between the United States and Europe, which grew from 507,000 in 1953 to 6,776,000 in 1969.[105] Instead, American would soon have to partner with foreign airlines to attract international passengers to its domestic routes.

Domestic Expansion

While the sale of AOA resulted in a contraction of the overall American portfolio, throughout most of the 1950s and 1960s, Smith attempted to expand American's domestic operations. He continued to lobby for new routes from the CAB though many of these were denied as the CAB sought to support smaller, regional carriers. American's size and scope became a detriment to further expansion of routes. Only 20 percent of American's passenger miles came from CAB awards since 1938, compared to 40 percent or more for most other carriers.[106]

Unable to expand at the rate and level that he hoped, Smith made a bold move to acquire Eastern Airlines in 1962. Its stalwart competitor, United, had acquired Capital Airlines in 1961, and, in so doing, it supplanted American as the largest U.S. airline carrier. American had also considered acquiring Capital but its heavy concentration of routes in the Northeast offered little real expansion opportunity. In contrast, Capital's routes provided mostly new territory for United. Smith decided not to contest United's acquisition of Capital and believed that United would support American's bid for Eastern. Smith also reasoned that Eastern's heavy concentration of routes in the Southeast would complement American's dominance in the Southwest and Eastern seaboard. Whatever goodwill Smith believed he had secured in walking away from Capital was all but nonexistent when he proposed the acquisition of Eastern.

American faced a hostile CAB, which feared that the combination of the second and fourth largest air carriers in the United States would stymie competition.[107] Resistance to the proposed merger was also strong among employees, union officials, and Congress—all fearing a loss of jobs. At the center of the resistance movement was C. E. Woolman of Delta. United also joined the chorus as a vocal and ardent critic of the proposed transaction. Though Smith fought to persuade CAB to approve the merger, he decided to withdraw his proposal instead of facing almost certain rejection.[108]

Despite the setbacks in expanding American's routes, Smith continued to make aggressive bets on new aircraft that would drive passenger-carrying economics. His pioneering endorsement of the DC-6 was followed by the adoption of the DC-7 (with more powerful engines and a cruising speed of 360 miles per hour). Smith purchased the DC-7 well ahead of the competition, and, in so doing, he secured the aircraft at a much lower rate. He used them to fly the first roundtrip, nonstop flights across the continent.[109] The DC-7 purchase was followed by the acquisition of the Boeing 707 jets in the 1950s. The Boeing 707 was the only plane he did not buy before his competitors, allowing Pan American and United to make their choices first between Boeing and Douglas so he could maximize his negotiating leverage. With Douglas getting orders from both airlines and Boeing from only one, Smith was able to get a bigger plane at a lower price from Boeing.[110] In choosing Boeing, Smith favored his company's needs over his own emotional attachment to an old business partner. It marked the first time that Smith "picked another manufacturer over Douglas" as the provider for American's flagship fleet.[111] In a cover story on the emergence of jets in the United States, *Time* magazine noted: "American's role in introducing the U.S. public to the jet age will be greater than any other line's. It carries 8,000,000 passengers per year, one in every six Americans who fly in the U.S., and almost twice as many revenue passengers as all overseas U.S. airlines combined."[112]

The new jets had a similar introductory fate as the DC-6s. Within the first year of their introduction, the Lockheed Electra (an early turboprop jet) was involved in a number of critical accidents that forced the Federal Aviation Administration (the successor to the CAB) to impose speed restrictions on jetliners until the problem (severe vibrations that led to structural weaknesses on the wings) could be identified.[113] Once the problems with the early jets were isolated, they also performed to early expectations. The Boeing 707 quickly outpaced the DC-7 and DC-6 in terms of speed and productivity. One Boeing 707 could do the work of at least two DC-7s; it could make two and a half transcontinental trips in the same time that the DC-7 made one cross country trip.[114]

During this time, Smith continued to push the envelope on the safety agenda that he initiated at the start of his presidency of the airline. He approved the funding for the development of a centralized crew training center that included state-of-the-art flight simulation equipment that operated round the clock. This effort resulted in the development of the American Airlines Flight Academy in Fort Worth, Texas that opened in 1971 and remains one of the most advanced pilot training centers in the world.

Twilight of the Patriarch

In early 1968 (the same year as Juan Trippe), Smith retired as chairman and chief executive officer, only to reemerge immediately as secretary of commerce in Texan Lyndon Johnson's cabinet, filling the post for the remaining months of President Johnson's term. In 1973, the 74-year-old Smith returned briefly to the helm of American to steady operations after a number of very difficult years. Smith's hand-chosen successor in 1968, George Spater, resigned under allegations of illegal campaign contributions to President Richard M. Nixon's reelection campaign in 1973. Though the scandal precipitated the early resignation of Spater, his tenure was fraught with a number of misguided management decisions and Smith's return was "like a dose of insulin to a diabetic. His mere presence helped to revive the sagging morale of American's workers."[115] When a new successor was named in late 1974, Smith walked away for good. As he was leaving American Airlines on his last day, Smith mentioned that he would never come back. When he was pressed to explain, he simply said, "If I start coming over here, and you guys ask my opinion, you have to remember I'm still thinking with a DC-6 mind and this business has changed. Yet, if you don't take my advice, I'll get upset. . . . [A]n old man should know when to quit. And that's why I'm never coming back."[116]

CHAPTER 6

William "Pat" Patterson and United Air Lines

In 1953, Selig Altschul, a prominent writer and consultant on aviation, cited one main reason for the persistent strength of United Air Lines in the airline industry since the early 1930s: "United was probably the first company to bring full-scale business methods and organization to air transportation."[1] Some observers of the aviation industry have minimized United's success because of its conservative business methods, but it is difficult to argue with the company's results over the long-haul: "United... consistently held to about 20% of the total passenger revenues generated by all trunk *and* local service airlines."[2] Having been selected as one of the primary air carriers during the tumultuous 1930s shake-out of the airline industry, United and its CEO, William "Pat" Patterson, chose to take a cautious and deliberate process of securing its dominance in the industry. During the second phase of the airline industry's evolution, United, like its counterpart American Airlines, sought ways to establish and protect the dominant business model that fit within the confines of government regulation.

Patterson was comfortable operating in the relative stability that regulation provided and took a very pragmatic approach to running his company within the established parameters. Many of the innovations that he championed to distinguish United from other airlines quickly became competitive imperatives that evolved into industry standards. Reflecting on his almost four decade tenure at United, Patterson commented:

> Although I am associated with an industry that's been government-regulated for many years, I wear no shackles. Our company and the other trunklines are self-sufficient... We have some requirements far more demanding than in other types of business. One of the pitfalls to be sidestepped in a regulated industry is the tendency to blame our troubles on the government. Grousing about the loss of liberty will not cause restrictive government agencies to go away... Trace the

> events that led to the origin of the restrictive agencies and you'll invariably find that a few men, or a group of men, chose to interpret liberty as the right to do as they pleased without regard to ethics or the interests of others. Their operations antagonized the public and Congress took action. It's as simple as that.[3]

As the airline industry grew and matured under regulation, it became increasingly important for large airlines like United to establish and reinforce the benchmarks for success, including operational and cost efficiency, economies of scale, and incremental innovations.

Patterson's strongest characteristic as a CEO was his unfailing commitment to plan for the company's long-term needs. He fearlessly absorbed short-term losses—either in the company's profits or market share—and argued to sometimes-wary stockholders that he was choosing the best options available on the path to sustainable growth. And although he always paid attention to the bottom line, he also was able to maximize performance through his attention to employee morale. His finesse lay in his ability to balance these two goals. During tough times that might have moved other CEOs to make cuts in personnel, salaries, or benefits, Patterson did his utmost to retain employees and shore up their loyalty to the company. Patterson commented:

> When anyone asks me what I consider my first responsibility to be and to whom, I say the public, the customer. And if I am asked to name the second, I say the employee. And then, third, the stockholder. In that order. Now let us see how logical that is. Without the customer you will not be in business. And without the employee to execute the great plans you have, some good, some bad, the plans might as well never be made. In the final analysis, this philosophy benefits the stockholder.[4]

Patterson's unwavering attention to maximizing efficiency and performance *no matter what* relieved him from the onerous task of major contraction of the company and the painful personnel decisions that contraction would entail. Keeping hold of personnel also strengthened United's corporate memory, an intangible but strong asset that Patterson kept alive through his policy of appointing most executives from within the company's ranks.[5] This may explain why Patterson chafed at the press's tendency to label him as paternalistic. As Patterson explained, "I have always had a horrible distaste for paternalism. I have tried to be thoughtful and helpful and my rule has been to have our personnel people inquire if anyone in trouble needed and wanted help, and give it if they did. But I never wanted to become a mother hen to United employees."[6] In short, maintaining good relations with labor was simply a tool of good management, especially when one believed, as Patterson did, that a firm's employees were its greatest asset. Patterson's management philosophy became institutionalized in the company "through a system of benefits and training, administered by one of the industry's first personnel departments. The absence of major labor difficulties during [periods] of union unrest testifies to the success of his efforts."[7] During his tenure at the helm of United, the airline was also the first to offer minimum wages, health benefits, and retirement funds for pilots.[8]

Patterson was adept not only at keeping the company lean when times were bad, but was also able to maximize profits when times were good. Perhaps most importantly, Patterson's long personal experience with economic hardship seems to have steeled him for the financial ups and downs United Air Lines would face in its early years. His strength under pressure helped him to recover from and then learn from his mistakes.

A Businessman First

One factor that distinguished Patterson from many of his peers in the field of aviation during its early days was his extensive experience in business. Before becoming a major player in the field of aviation, Patterson had relatively little experience with flying machines. CEOs like Woolman, Rickenbacker, and Trippe along with aviation leaders such as Guggenheim could all have been considered among the very first generation of American pilots. Born in Hawaii in 1899, Patterson's life started out without the money, opportunities, or even the free time to entertain notions of flying. Although Patterson began life as the son of a relatively successful manager of a Hawaiian sugar plantation, his father succumbed to malaria when Patterson was only seven years old. From that time on, Patterson and his mother were both forced to focus on making money to survive. Unable to eek out a decent living in Hawaii, Patterson's mother Mary (with financial help from her father) moved to San Francisco to attend a business school for six years while her son was enrolled in a military academy in Honolulu. Heartsick to be separated from his mother, Patterson escaped from the academy and went to San Francisco as a cabin boy on the *Annie Johnson* at the tender age of 13.[9]

After graduating from a local grammar school at age 14, Patterson looked immediately for work. Brandishing a scholarship medal earned from his good grades, Patterson convinced a cashier at Wells Fargo Bank to hire him as a messenger with a monthly salary of $25. Early on, he profited from the sound recommendations of others, following the counsel of the very cashier who hired him, an Australian immigrant named F. I. Raymond. Learning that Patterson was not planning to go to high school, Raymond invited the youngster to dinner and passed on this advice: "If you want to get ahead, you'll have to keep on studying.... People will never criticize you for not having an education, but they will if you don't try to get one."[10] Patterson was convinced and began a 13-year education in night school, eventually earning 6 semesters of university credits from the American Institute of Banking.[11] After his mother remarried when he was 15 years old, Patterson's employer Wells Fargo became a veritable home away from home when he moved in with a dozen or so fellow employees from the bank (mostly college graduates) into a boarding house. Eager to succeed as well as to make friends to fill the financial and emotional voids created by his tumultuous family life during this time, Patterson combined hard work and superior social skills to slowly climb up the ladder at the bank.[12]

By the time he was 27, Patterson's salary had risen to $350 a month as he assumed the position of assistant to a vice president. Although his success had been steady, his late teens and early twenties had still been hard. Patterson recalled how his poverty shaped these years: "I was often hungry. I could spend only 25 cents for a meal. At a Boos Brothers cafeteria on Powell Street I always ordered the same things, for fear that any variation might cost a nickel or more."[13] But lessons learned from the poverty and uncertainty that afflicted his young life proved to be of great value to Patterson as a CEO. Patterson's mentor Raymond believed that the best way to ensure one's future was constant self-improvement. "If I had walked into the arms of a union organizer after that first day at Wells Fargo," Patterson mused, "instead of into F. I. Raymond, I might have sought security as an office boy.... Instead, I had all the advantage of complete insecurity."[14] Deeply ingrained in Patterson's psyche was the knowledge that no matter how secure or permanent something might appear—such as a father's love for a young boy—disaster could unsettle anyone at anytime. But great pain had also gifted Patterson with the power of empathy. Reflecting on his early years, Patterson explained: "When I look back at those early days, which were pretty tough, I feel sorry for the youngsters of today who miss those experiences.... You get something out of them. You learn to know what's going through the hearts and minds of people in trouble. I can live their troubles with them."[15]

Although hardship had made Patterson careful and meticulous, it did not make him timid. At the age of 19, Patterson paid $5 (25 percent of his monthly salary in 1914) to take a ride in a plane with a barnstormer who had stopped for a few days in San Francisco. Inspired by his first flying experience, Patterson soon afterward was preparing to fly again when, just as his name was being called to take to the sky, another barnstormer performing stunts crashed to the ground. The dangers of airplanes could not have been made any clearer, but Patterson remained enthusiastic about the potential of flying machines in the future.[16]

About eight years later, Patterson began his career in the airline industry in an unexpected way. The president of a struggling local airline, Pacific Air Transport, came into Patterson's Well Fargo office in March 1927 while most of Patterson's colleagues were out to lunch. The president, Vern C. Gorst, was aching for Pacific Air to become the first airline to win an airmail contract with the post office; unfortunately, he had little money to buy the new planes that he needed. Intrigued by the prospects of Pacific Air, Patterson went to the airline's offices to examine the business and its books. Then, without formal permission from his superiors, Patterson loaned Gorst $5,000.

Surprised by Patterson's unusually impulsive decision, Wells Fargo President Frederick Lipman was worried that the loan might have been a bad decision that would undercut Patterson's confidence and future performance with the bank. Lipman encouraged Patterson to safeguard the loan by overseeing Pacific Air's finances with some care until the money was repaid. Patterson jumped into the role with gusto and became an unofficial but influential financial advisor to Pacific Air. His contributions to Pacific Air culminated at the end of the year when the still struggling Gorst was looking for a buyer to keep his airline afloat. After receiving two offers, Patterson convinced Gorst to accept the proposal to merge with Boeing Air Transport (BAT), even though the final purchase price for Pacific Air was less than the one offered by a competitor, Western Air Express. Patterson preferred the deal with Boeing because it included a provision that would keep on all of Pacific Air's employees. In January 1928, BAT bought 73 percent of Pacific Air's stock.[17]

BAT was a part of a large aviation conglomerate hoping to grow even larger. BAT's origins stretched back to 1917, the year the Boeing Airplane Company (BAC) was founded. Ten years later, Boeing's airplane division won a postal contract to deliver mail between Chicago and San Francisco and then created BAT to fulfill the contract by using Boeing's own B.40-A airplanes. Soon afterward, the two Boeing companies merged on October 31, 1928 to form the Boeing Airplane and Transport Company—quickly renamed as United Aircraft and Transport Corporation (February 1929), which brought together a number of manufacturing companies, most notably the airplane engine manufacturer Pratt and Whitney.[18] Hungry for talent to manage what was, at the time, "one of the largest aeronautical organizations in the world," Phillip G. Johnson, then the president of United Aircraft, had been impressed with Patterson's competence as well as his stewardship of Pacific Air's employees and decided to offer him a position as his assistant in the winter of 1929. Deeply flattered by this unexpected promotion and attracted by the excitement promised by the growing airline industry, Patterson accepted.[19]

Joining United Aircraft and Transport Corporation

Arriving at Boeing's headquarters in Seattle just two months after the formation of United Aircraft and Transport Corporation, Patterson brought business skills along with a penchant for efficiency that soon made an impact on an industry shaped by adventuresome pilots. His first responsibilities were to rationalize the business practices of Pacific Air Transport and then to systematize the operations of the Boeing airplane factory as well as for BAT itself. An historian of United Air Lines described Patterson's role in the Boeing conglomerate: "Largely because nobody else topside thought much about it, Patterson became the airline's policy man and planner."[20]

Patterson found his responsibilities began to grow beyond those of an assistant. Patterson's boss, Johnson, was stretched running the holding company and relied on Patterson to work out the details of the merger between BAT and Pacific Air Transport. While Patterson focused on the airlines in the west, the vision of a transcontinental airline now attracted the attention of United Aircraft's directors. At a dizzying rate, United soon added airlines to its portfolio to build a transcontinental route: Stout Airlines, connecting Chicago to Detroit and Cleveland; National Air Transport, which ran routes between New York and Chicago; and Varney Air Lines whose routes ran from Nevada to Seattle.[21]

At the completion of the purchase of Varney Air Lines in 1929, all four airlines that were part of the United Aircraft holding company acted independently of one another. That situation began to change when the president of BAT died suddenly in the winter. Johnson responded by making Patterson the head of BAT. In addition, Johnson then decided to move toward integrating the airlines owned by United Aircraft by making Patterson vice president of all four of them while Johnson moved to New York to devote his attention to the overall holding company, United Aircraft and Transport Corporation (while he still retained the titular presidency of United's four airlines).[22]

Although Patterson was not yet officially president of any of the airlines he helped to supervise around 1930, his style of management, which drew heavily upon the collegial atmosphere he enjoyed with Wells Fargo employees (whether at the office or in his boarding house), gave him an early opportunity to implement an experiment in airline service that would have a huge impact on Boeing's growing airline empire as well as the entire airline industry. That experiment began with an idea from a nurse who worked at San Francisco's French Hospital, Ellen Church.

Early Innovations in Air Service

In 1930, Boeing's district manager for San Francisco, Steve Stimpson, was returning home on a flight from Reno, Nevada when the plane ran into bad weather. Normally, the pilot and co-pilot flying Boeing's planes were also in charge of cabin service and served coffee and sandwiches to the 10 to 20 passengers who filled the small planes used during the late 1920s. Preoccupied with getting the plane through this rough patch, the pilots were grateful that Stimpson volunteered to step in and take over the in-flight service. When Stimpson finally got back to his office, he suggested that Boeing think about adding a male steward to the flight crew—perhaps a teenage boy would do (which might add a helping hand with minimal additional weight).[23] Coincidentally, just after Stimpson mailed in this proposal, Ellen Church took the initiative to walk into Stimpson's office with a more novel idea. Church was a flying enthusiast but saw that the possibilities for women to enter the piloting profession in the 1920s were severely limited. Her solution: Why not employ nurses to tend to passengers? The presence of female stewards/nurses might reassure passengers of the safety of flying; they could also provide medical help for passengers in cases of great discomfort or emergency.

Stimpson loved the idea and forwarded it to Boeing's director of airline traffic, who responded with a terse, "No." Undoubtedly, the Boeing executive shared the skepticism of many pilots who were worried that "frail" women stewards would prove to be more of a burden to the pilots than an asset for the passengers. Undaunted, Stimpson tried to persuade Patterson (who was the airline traffic director's boss) with several messages that conveyed his enthusiasm. In his first memo, Stimpson claimed: "Imagine the psychology of having young women as regular members of the crew. Imagine the national publicity we could get from it, and the tremendous effect it would have on the traveling public. Also imagine the value they would be to us not only in the neater and nicer method of serving food but in looking out for the passengers' welfare."[24]

Unlike many at BAT, Patterson seemed open-minded enough to give the idea some thought. Like his contemporary Harry Guggenheim, Patterson was quite aware that the biggest obstacle to growing the airline industry was the perception that flying was not a very safe activity. Certainly, any service that might help to enhance passengers' perception of flying as a safe and comfortable experience would be a plus. In addition to talking the idea over with colleagues at the office, Patterson consulted his wife Vera who wholeheartedly supported Stimpson's initiative. Not long afterward, Patterson gave Stimpson and Church permission to begin on a small, experimental basis. Other airlines began to follow suit and stewardesses soon became so connected in the public's mind with quality service in transportation that bus lines and railroads experimented with stewardesses briefly.[25] As early as 1932, the *New York Times* made this remark: "The lanes of the air have been made smooth for many a passenger by a stewardess, co-pilot, or hostess."[26]

Patterson's willingness to entertain ideas from employees at any level of the corporation allowed a promising idea to percolate up from the ranks that eventually shaped the entire airline industry. Open communication under Patterson's leadership was one of the distinguishing characteristics of BAT and, later, United Air Lines.[27] By the mid-1930s, Patterson had institutionalized this bottom-up cultivation of ideas through the establishment of a monthly "Suggestion Conference" in which six executives and six "rank-and-file" employees met monthly to discuss suggestions for improvements in any part of the airline. Over the next 20 years, more than 200,000 suggestions were fielded by this group saving, according to Patterson,

Quarter century of flight fashions: Uniforms worn in the 25 years since "sky girls" first went aloft are modeled by United Air Lines stewardesses. They are (from left) Carol Roos, who sports the original 1930 outfit; Carol L. Smith, 1933; Ruth Warren, 1936; Connie Ammon, 1937; Norma Banks, 1939; Nancy Riley, 1941 and Aldys Holmes, 1955. (Source: Underwood & Underwood/CORBIS).

millions of dollars. "But that's just the dollar side of it," Patterson pointed out years later. "The suggestion conference has been worth even more in morale. It keeps everybody on his toes, trying to run the airline better."[28]

Patterson "thrived on change. He... picked other people's minds for their brainstorms, which he delighted in putting into practice. Most empire builders, when they get their empires molded, want to keep them that way. Patterson [did] not. He... liked his empire of the air in a state of flux."[29] For a trial-and-error approach to be effective, Patterson would have to be willing to abandon experiments quickly that showed little promise for success. One short-lived idea, coming again from Steve Stimpson, was to get more women to fly by permitting businessmen to bring along their wives for free. The program was abandoned soon after Patterson learned that many businessmen taking advantage of this program were accompanied by women other than their wives.[30] Similarly, Patterson introduced a "Sky Lounge" program that was meant to provide extra comfort for passengers; it was terminated when passengers did not agree that a roomier plane (14 instead of 21 seats) was not worth the extra $2 in airfare.[31]

Notwithstanding these failures, experimentation led to more groundbreaking developments later in the 1930s that would grow exponentially under Patterson's guidance. One particularly fruitful experiment came from Patterson's own initiative when he felt increasing dissatisfaction with the airline food he ate on his many trips on United. After discussing the problem with a professional in hotel management in San Francisco, Patterson learned of a food-service expert named Don F. Magarrell. Soon afterward, Magarrell suggested more efficient placement of food galleys within United's airplanes and then lobbied to establish United's own food kitchens. The airline's kitchens, located at several airports, would replace the airline's use of caterers whose food became cold and unappetizing while being transported from adjacent city centers. After an initial investment of $3,000 to establish an experimental air kitchen, manned by expert Swiss chefs in Oakland, California, the new food soon became a hit with passengers in 1936.[32] It was, according to one critic, "the greatest aviation advance in recent years, as far as women's comfort is concerned."[33] Other airlines would take years to follow suit, such as Continental, which did not begin its own food kitchens until 1946.[34]

In the regulated environment of the airline industry, there were relatively few ways in which airlines could distinguish their businesses. Providing innovative services such as medically trained stewardesses was one way to build added credibility and brand equity in the marketplace. It also had the added benefit of contributing to a shift in the perception of airlines as dangerous; stewardesses offered a sense of comfort and familiarity in an otherwise frightening and unknown endeavor. Many of these services such as stewardesses and flight kitchens at United and safety promotions and faster aircraft at American Airlines became engrained in the overall airline industry resulting in an upward shift in the requirements and context for long-term success. These services were quickly absorbed into the dominant business model, shifting them from

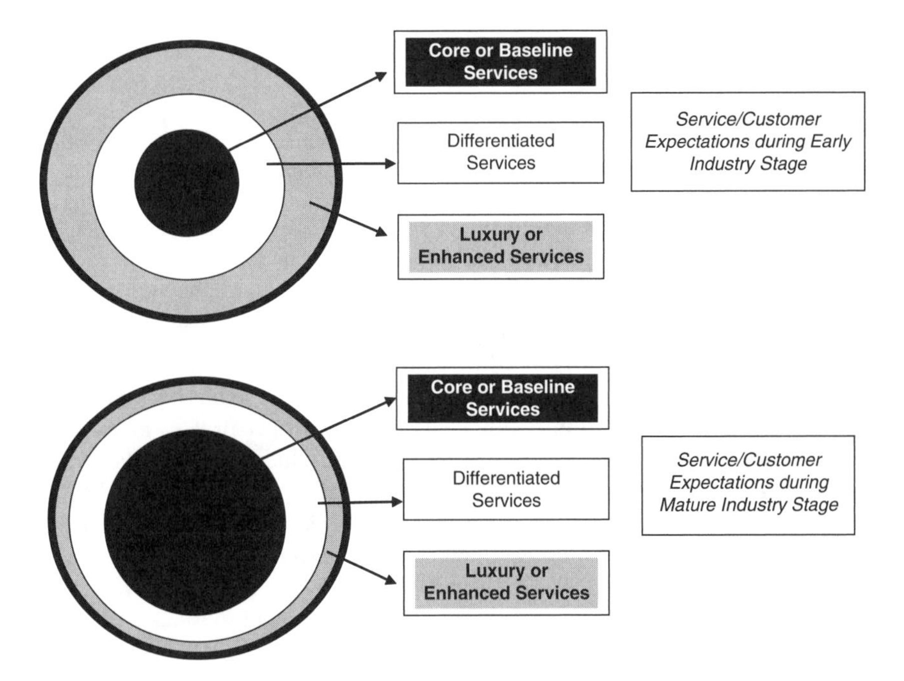

Figure 6.1 Core services expand as industry matures

distinguishing luxuries to base necessities, which expanded the core (see figure 6.1).[35] When these services became part of the core offering, the entire industry was forced to adapt.

Impact of Early Airline Industry Regulation

Despite these successful and influential experimentations, United Air Lines went through difficult times in the 1930s, as did all airlines across the industry. In the immediate aftermath of Senator Hugo Black's public investigation of the airlines under the regime of Postmaster Brown, United lost all of its airmail contracts. In accordance with the provisions of the Air Mail Act of 1934, United Air Lines had to be broken off from its parent, United Aircraft and Transport, and its president, Philip G. Johnson, who had attended the spoils conference in 1930 with Postmaster Brown, was forced to resign from the airline. With this shift, Patterson was thrust into the presidency of United Air Lines in 1934. When he assumed the new role, "he traveled the length and breadth of United routes, meeting each of his 1,400 employees in order to understand their concerns and problems."[36] By that summer, the new airline emerged out of the reorganization with no debt and approximately $4,000,000 in working capital.[37]

When Roosevelt retracted his disastrous decision to transfer airmail responsibilities to the army and reopened bidding from private carriers to carry the mail just a few months after they had been cancelled, Patterson faced a difficult decision. Reflecting on his attempt to regain the routes United had lost, Patterson described his dilemma:

> We were losing so much money [because of cancelled airmail contracts] that we had to have some relief. We had two courses to follow. One was to be sure that you would get your route back and be ridiculous in your bid, and the other was trying to use good judgment and bid the lowest you possibly could but no lower than your previous experience indicated would be ridiculous and something that could not be substantiated to your stockholders. That is the way I felt about it.[38]

As it turned out, United, propped by its reputation for good service and good management, was able to regain all of its routes except for one linking Chicago to Dallas. Braniff secured this route with a bid of 22.5 cents per airplane mile, which was 17 cents less than United's bid.[39] Interestingly, the Chicago to Dallas route was the only one in which United faced competition. Patterson's gamble to bid for United's previous routes at rates that were only slightly lower than it received before 1934 was well founded. This strategy, however, did not work when Patterson sought brand new routes for United. The company was underbid on all of the new routes.[40] Other airlines that secured new routes or reclaimed their mail subsidies with lower bids than United were stuck with far lower revenues. (Compare United's average airmail contract rate of 39 cents per airplane mile to American Airlines' average of 25.3 cents.)[41] Despite his best efforts to garner more money for airmail service, Patterson struggled to produce a profit for United. United like almost all the other airlines lost money on airmail service immediately after the rebidding process.[42]

"Looking back," commented Patterson, "the air mail cancellations, although a terrific jolt to the companies and unfair to individuals, were a blessing in disguise.... It was a spanking that made us better boys."[43] That spanking resulted in a slow but very determined strategy by Patterson to move away from depending on revenues from airmail while making a simultaneous effort to expand profits from passenger and cargo service. United's employees responded to Patterson's call to expand passenger service. In 1933, United logged 69 million passenger miles that represented 40 percent of the company's income; in 1936, United flew more than 100 million passenger miles that brought in 58 percent of United's earnings. United's cargo service, called "air express," although modest, more than tripled its earnings during the same period from $133,000 to $431,000. Yet even with these gains, United's losses were eating into its initial capitalization, and its profits declined from $2,147,000 in 1930 to $371,000 in 1936.[44]

As deft and flexible as Patterson was, he did operate United Air Lines during this critical period with a significant, self-inflicted handicap: deep loyalty to his former boss, Philip G. Johnson. Although Johnson's forced departure from United did not mark the end of Johnson's career—he immediately founded an airline in Canada in 1937 and returned to Seattle to run Boeing in 1939—Patterson never forgave the U.S. government for what he believed was an arbitrary and excessive punishment for having attended the spoils conference of 1930.[45] In 1934, Patterson began a legal proceeding against the federal government in an effort to clear Johnson's name. One of Patterson's biographers describes the details of the suit: "About the only means of vindication, except by public opinion, was to bring suit in federal courts. The United States Supreme Court refused to consider a suit to clear Johnson of collusion charges. But the U.S. Court of Claims accepted a suit by PAT and BAT to recover $3,100,555.43 in postal revenues allegedly lost by the arbitrary airmail cancellation. Winning this suit, to Patterson's way of thinking, would imply vindication for Johnson. It dragged out for years, and became known as the 'long suit.' A token award to United was finally handed down in 1943."[46]

At the same time that Patterson was initiating the "long suit," United was petitioning the federal government through the post office for crucial extensions of its airline service that were needed to expand its operations. Many of United's petitions were refused by the post office. Postal officials friendly to Patterson quietly made it understood that the "long suit" was becoming an obstacle to United's growth: "we could see United's problems more clearly," one postal representative explained, "if smoke didn't get in our eyes."[47] Undaunted, Patterson tried to acquire smaller airlines to achieve United's plans for growth; most of these petitions were also denied.[48]

Patterson's stubbornness had real consequences. Boxed in by the post office—who granted competitors American Airlines and TWA another 60 routes immediately after 1934—United's share of revenue passenger miles in the United States plummeted from 44.8 percent in 1934 to 23.3 percent in 1937.[49] By that year, American Airlines rose to take the prize of being the largest airline in the United States. Long known as #1 in the airline industry, United fell with a thud to #2.[50]

Similar to other executives in the airline industry, Patterson sought to differentiate his airline with new state-of-the-art equipment. Patterson actually inherited this strategy from United's former holding company United Aircraft and Transport. In 1932, the Boeing Company received an order from United Aircraft for 59 Boeing 247s that would be used by United Aircraft's 4 airlines. At the time it was delivered in February 1933, the Boeing 247 was the best commercial aircraft around. Its Pratt and Whitney engines and its all-metal frame made the more flimsy trimotor airplanes a thing of the past.[51] It had a cruising speed of 160 miles per hour and could hold 10 passengers. Boeing proudly announced in 1933 that this innovative and powerful airplane would not be available to airlines competing with United Aircraft's airlines.

Boeing's snub of the rest of the airline industry enraged TWA in particular and motivated the airline to send a letter to a handful of other manufactures announcing that it wanted to make an order for an airplane that could easily outperform the Boeing 247.[52] Douglas Aircraft answered the challenge and created the 14-seater DC-2 in 1934 that could cruise at 196 miles an hour.[53] Airline historian R. E. G. Davies notes, "Within a span of eight days [of its introduction], it [DC-2] broke the speed record from New York (Newark) four times, and virtually chased the Boeings off the route, knocking half an hour off the 247's 5 ½-hour flight time."[54] Not wanting to be outdone, American Airlines soon placed a significant order for what would become the DC-3, an aircraft that could cruise at a similar rate as the DC-2 but could hold seven more passengers. As discussed in the previous chapter, American Airlines' CEO C. R. Smith ordered 20, with an option to buy another 20 in the future.[55] This was the plane that would finally make it economically feasible to run an airline exclusively for passenger transport.[56] Suddenly, the much-heralded capabilities of the 247 hung like an albatross around United Air Lines' neck, with passengers clearly preferring the "larger, quieter, and faster DC-3's" that American put into service in 1936.[57]

With the new DC-3 menacing to eat into United's meager profits in the middle 1930s, Patterson had two choices: improve the 247s or convert to the DC-3s or an equivalent. At first, when TWA started to use DC-2s, Patterson played it safe and decided to invest $1,000,000 in improving the performance of the 247s. By improving the engines, United made the 247s go 10 miles an hour faster—a change that hardly made waves in the industry. The 247 suffered more shame when it lost a highly publicized London to Melbourne air competition. Other attempts to draw passengers to the 247 through advertising also failed.[58] Immediately following the DC-3's arrival, TWA and American Airlines, began to make money with the new plane "as fast as the half-filled Boeing's lost it."[59] Patterson finally relented and ordered 10 DC-3s for $1,000,000 for delivery in 1936.

Despite the problems derived from Patterson's stubborn loyalty to Johnson—as well as to the Boeing 247—Patterson came up with a timely plan that promised to salvage the legacy of United Air Transport's investment in the Boeing airplane (along with Patterson's million-dollar upgrade) as well as to compensate for United's unsuccessful attempts to expand its operations during the 1930s. Patterson described his vision to United's board of directors in December 1935. The impetus for the plan could be traced back to the dark days of 1934, when "the fall-off in passenger business" resulted in "a surplus of at least 12 airplanes."[60]

Patterson explained to the board that United's formerly dominant position in providing transcontinental air service was deeply impacted by TWA's faster, nonstop service (see sidebar).[61] With this new challenge to United's market share, "it seemed logical to build up certain feeder lines into productive territory with more modern equipment in the hope that this would result in increased short haul business of feeder lines into United to offset the loss to TWA on the long haul business." Building up these feeder lines would entail leasing the surplus Boeing 247s without profit (often with an option to buy) to four feeder lines located in the West and the East of the United States.[62] The airlines and the territories they served were (1) National Parks Airways, operating between Salt Lake City and Great Falls, Montana; (2) Western Air Express, operating between Salt Lake City and San Diego via Los Angeles; (3) Pennsylvania Airlines & Transport Company, operating between Milwaukee and Washington via Detroit; and (4) Wyoming Air Service, operating between Cheyenne and Pueblo, Colorado via Denver. Agreements with these four airlines were made in late 1934 and early 1935.[63]

TWA's Jack Frye and the Lockheed Constellation

In the early 1930s, Jack Frye of TWA and Patterson of United were involved in developing pioneering airplane technology—the Boeing 247 for United and TWA's response, the Douglas DC-2. In this early round of technological competition, Patterson's loss moved him to become more conservative in depending on technology. Jack Frye learned a very different lesson, which had important consequences for TWA.

Frye, unlike Patterson, was an experienced pilot—one might even say a daredevil, starting off his career doing stunt flights for movies, including *Wings* by Howard Hughes in 1927. Ten years later, Hughes saved TWA from its financial woes by buying the airline and running it in a partnership with Frye. Hughes and Frye shared the same tendencies: a fascination with airplane speed and performance. Both men wanted to make a technological coup to shake up the industry. "Above the weather" flying was a particularly important objective for Frye who had been interested since 1939 in acquiring a plane that could at least go higher than 20,000 feet, which would avoid 80 percent of the bad weather below.

The answer was the Lockheed Constellation. After being developed behind an iron-curtain of secrecy to keep the rest of the airline industry in the dark, the Lockheed-049 Constellation made its debut in 1944. In April of that year, Frye along with fellow daredevil Hughes made the first coast-to-coast flight of an airliner in the amazing time of just less than seven hours. The previous record of 10 hours and 22 minutes had been set in 1935. The publicity the trip garnered was great, and Hughes—planning to lure celebrities to publicize the virtues of the Constellation—hoped it would be the opening salvo in drawing more passengers to TWA.

The short-term rewards of the Constellation were great because the airplane clearly outperformed all others on the market. However, Frye's eagerness to place most of his airline's eggs in the Constellation's basket was a great risk—one that United's Patterson had shunned since the days of the Boeing 247. In another contrast to Patterson, Frye—described as "brash, energetic, and fiercely competitive"—shunned the counsel of TWA's technical experts, who believed that the plane was too complex and, while aesthetically appealing, its sleek design (certainly attractive to pilots) actually decreased the number of passengers it could carry.

The imagined competitive advantages of the Constellation quickly plummeted in just a few short years. In 1946, five TWA employees were killed during training when an electrical fire, caused by a "small 'lead-through stud' at the point where the wing is joined to the fuselage," forced a Constellation to crash land. This accident led to the temporary grounding of TWA's 11 Constellations. Frye blamed subsequent financial losses by TWA on the grounding.

This mishap, along with a pilot's strike, allowed Howard Hughes' assistant, Noah Dietrich, to lead a revolt to expel Frye from the company. Before he could be ousted, Frye resigned in February 1947. Although newer versions of the Constellation would fly again, the airplane proved to be the undoing of Frye's tenure at TWA.

With this strategy, Patterson offset the losses incurred due to the "long suit" by drawing more passengers to United Air Lines thanks to the help he gave to smaller airlines that connected with United's transcontinental route. The authors of the *Corporate and Legal History of United Air Lines*, writing in 1953, reflect how the decision to lease the Boeing 247s to feeder airlines 18 years earlier became a pivotal moment in the development of this business ethos within United: "Although United was undertaking to develop business from the connecting airlines concerned, they also benefited materially from the above transactions. For the most part all [the connecting airlines] were sustaining substantial losses, their financial resources were low, and their prior airplanes were of old types. Thus, in a time of need, United made available to those carriers modern equipment, and, by reason of favorable charges in terms of payment, gave them considerable financial assistance. At the same time and as a result they were enabled to improve their services to the public."[64]

A crucially important lesson Patterson learned from this experience was the benefits United gained from the growth and health of the entire airline industry. This became a hallmark of Patterson's leadership in aviation; he befriended CEOs from major competitors—such as Robert Six from Continental and C. R. Smith from American Airlines—in the belief that United's profits could be generated along with, rather than despite, the success of other airlines. In the 1960s, Patterson explained his reasoning: "Sometimes I get mad at our competitors for the silly things they do, and they get mad at us for some of the things that we do. One man's idea is another man's challenge to compete. Air transportation wouldn't be where it is today if it hadn't been for some rugged individuals fighting for business, each in his own way."[65]

DC-4 United Mainliner Flying over San Francisco Bay. (Source: Douglas Aircraft photograph collection. Baker Library Historical Collections, Harvard Business School).

Just a few years afterward, Patterson abandoned the go-it-alone strategy in airplane technology. In the wake of well-publicized air accidents around 1937, United wanted to develop even safer planes than the DC-3 for the whole industry. Patterson convinced TWA, Eastern, American, and Pan American to chip in half of the development costs for a new Douglas prototype. Thanks to this initiative by Patterson, which was unique in the industry, the DC-4, a 4-engine airplane capable of carrying 40 passengers, was created in 1939.[66] Patterson placed orders for 20 DC-4s but the outbreak of hostilities in Europe forced Douglas Aircraft to cancel the order; aircraft production for the next five years was devoted exclusively for military purposes.[67]

The War Years

Although World War II might be seen as an anomalous time in business history when so many industries prospered thanks to the material needs for the U.S. military forces, the vastly new circumstances facing the airline industry also posed considerable risks for United as well as the airline industry as a whole. The first of these potential problems was President Roosevelt's initial decision to nationalize the airlines for the purposes of national defense.

For a short while, Roosevelt had every intention to bring the entire operations of the nation's civilian air services under the direction of the U.S. government.[68] Luckily, the president of the Airline Pilots Association, Edgar Gorrell, argued that the existing personnel and infrastructure of the commercial airlines could be adapted very quickly to the needs of the nation—certainly much more quickly than if the army or navy were suddenly charged with running the nation's airways. Roosevelt soon gave up the idea, opting instead to enlist a large portion of the commercial airliners for military needs and relegating the rest to fly reduced schedules for the airlines.

Patterson, formerly irked with the government for its treatment of Johnson, quickly made himself as useful and cooperative as possible. Immediately after learning at a meeting at the Department of Commerce in Washington, D.C. that the government was planning to acquire 36 DC-3s out of United's fleet of 69, Patterson enthusiastically answered the call of duty: "Gentlemen, I can settle United's part in this deal in ten minutes. What you want from United is perfectly all right with me. We'll deliver the planes to you as you need them."[69] In an internal memo to employees, Patterson made the stakes clear: to preserve the airline and the air transport system it is a part of, United's employees will have *"to do the things we are told to do and do them well. . . .* Furthermore, there has been no conversation of confiscation of the airlines and I am sure there will not be if we properly organize our jobs and maintain an outstanding patriotic viewpoint in the performance of our duties."[70] Unlike Juan Trippe, Patterson's demonstration of patriotic fervor kept the government happy, whereas Trippe's balking at government service during the war may have caused real problems for Pan Am's future. The satisfaction of the government with the performance of Patterson and much of the airline industry prompted a glowing comment from the chief of the Army's Air Transport Command,

General Harold L. George, just a few months after Roosevelt gave up the idea of nationalizing the airlines. The army would not be taking over the airlines, George declared; instead, "the air lines are taking over, taking over the biggest job they ever tackled, and we of the War Department have the utmost confidence that they can carry out the task."[71]

Cooperating with the war effort had the long-term benefit of keeping United in private hands along with the short-term benefit of rising profits. Although United's number of planes had fallen, they were almost always filled to capacity with military and civilian passengers, as well as airmail. Increased demand, combined with increased efficiency, led United's load factor (the total carrying capacity of its airplanes) to jump from 61 percent in 1939 to 96 percent in 1944 (with net income rising between those years from $776,000 to $6,614,000).[72] United's airmail and air cargo grew tremendously during the war, however, Patterson was not swayed that shifting United's business to carrying cargo was a prudent postwar strategy. Despite some rosy predictions for the continued growth of air transportation of goods around the world following the end of World War II, Patterson believed such predictions were influenced by the intoxicating—but very temporary—government largesse during the war. "If all the hot air on the subject circulating today were stored," Patterson maintained, "it would create enough energy to fly all the planes in the U.S. without gasoline."[73] By the mid-1940s, Patterson knew that air transport planes were not efficient enough to challenge traditional shipping in the cargo industry.[74] After the war, United returned its attention to passenger traffic.

Postwar Growth

As United's early history suggests, Patterson's success was multifaceted. He led the airline through maintaining high morale, keeping lines of communication open and ideas flowing at all levels of the company, demonstrating a willingness to admit mistakes, and making efficiency and planning central parts of United's approach to business. He also made important decisions to innovate (e.g., in incorporating flight kitchens) as well as decisions not to proceed down certain risky paths (in opting not to expand cargo service past the end of World War II). During this time, Patterson committed the company to develop an extensive research and communications laboratory focused on resolving persistent safety issues. Through the investment in what Patterson called the "Flying Lab," United developed radar-equipped aircraft, terrain clearance indicators, and two-way air-to-ground radio transmission.[75] Like some of United's other innovations, the work of the "Flying Lab" benefited the entire airline industry, once again changing the contextual landscape.

In the immediate postwar years, United's efficiency skyrocketed. Between 1946 and 1952, the production of each employee, as measured by ton-miles of payload, increased from 9,100 ton-miles to 21,000. Profits soared along with this efficiency from $1,000,000 to $10,000,000.[76] With United's strong suit being its transcontinental routes, large and reliable airplanes increased profits by transporting higher numbers of passengers across the country with fewer stops for refueling along the way. The two airplanes principally used by United in the years immediately following the war—the Convair 340 and the Douglas DC-6B—each had an annual seat-mile capacity that was approximately 10 times higher than that of the DC-3.[77] The Douglas DC-6 could seat 50 passengers and carry them more than 4 times farther than the DC-3—approximately 2,200 miles.[78] Revenue-passenger miles increased, too, during the early post-war years: from 1 billion in 1946 to 2.4 billion in 1952.[79]

United did not expand through efficiency alone; expansion through acquiring carefully chosen new routes also aided United's bottom line. One area where United decided not to enter into was transoceanic air travel. To the surprise (and dismay) of many airlines that wanted to compete with Pan Am in the international field, Patterson supported the idea of continuing to make Pan Am America's primary international airline in testimony before the Civil Aeronautics Board (CAB) in 1944. United's economic forecasters believed there would be insufficient worldwide demand to prompt another American airline to enter the international arena. Given the government subsidies foreign airlines enjoyed, Patterson added, "I feel very strongly that the only way for us to compete with foreign carriers is through a chosen instrument."[80]

Instead of flying to Europe or Asia, Patterson had his eyes on a far-away destination with important ties to the United States: Hawaii. Although Hawaii was Patterson's childhood home, establishing United's presence there was no act of nostalgia. In the postwar era, Hawaii promised to become a major tourist destination for many Americans. Even though Hawaii's hotel capacity around 1950 may have created a temporary impediment to the growth of air travel there, Patterson believed in the long-term potential of the route and succeeded in securing permission from the CAB to establish San Francisco-Hawaii service in 1947, followed by Los Angeles-Hawaii service in 1952. Pan Am was United's major competitor in this market, and Trippe's lack of experience in dealing with talented competitors created an opening for Patterson. Thanks to aggressive marketing in the United States, the number of passengers taking United to Hawaii surpassed Pan Am's totals as early as 1951: United flew 42,000 passengers, as opposed to Pan Am's 35,000. Even more importantly than these numbers was the larger impact of adding Hawaii to the United route structure. Adding this new destination encouraged more transcontinental traffic *en route* to Hawaii. United benefited from its extensive domestic route system that functioned as feeder lines for its transpacific flights. Even more traffic along this long east-west corridor was encouraged thanks to "interline arrangements" between United and foreign companies who were competing with Pan Am in the transatlantic airline market.[81]

One vivid example of how Patterson's careful planning yielded economic benefits while simultaneously maintaining United's *esprit de corps* came in the wake of the inauguration of United's airline service to Hawaii.[82] This immediately expanded United's routes from a little fewer than 3,000 miles in an east-west direction to more than 5,100 miles. Less than a year later, Patterson effectively decentralized the airline's operations into three different centers across the United States in order to better serve this sprawling route system.

These specialized centers of operations helped to prevent the development of autonomous corporate fiefdoms within the company because each center provided essential services for the whole airline. Open communication between those centers was further encouraged through an "extensive communications system consisting of private telephone and teleprinter circuits."[83]

The company's headquarters remained in Chicago while its operations hub moved to Denver and a new state-of-the-art maintenance facility was built in San Francisco. The headquarters in Chicago helped United maintain close ties to the financial and legal powerhouses of that city. Denver was at the geographic center of the United route system, and the operations center there monitored the daily progress of all of United's planes as well as the ancillary flight activities required to service and provision planes along their routes. San Francisco was home to a new kind of maintenance base where airplanes stopped annually to receive an efficiently designed overhaul in which United's planes were serviced in an assembly-line fashion. Before leaving San Francisco, all of the planes were equipped with the latest technological and mechanical improvements.[84] This maintenance facility gave United a strategic advantage through the company's "notably improved aircraft utilization and lower operating costs . . . relative to its major competitors." Far ahead of its competition, United's decision to invest $4.5 million in its maintenance facility gave the company, according to Selig Altschul, "a 'built-in' advantage which can hardly be equaled by any of its competitors except at a materially higher cost and in terms of years."[85] Here we see another way in which managers add value. They innovate internally in administrative structures as much as they innovate in the elaboration of external products and services. Patterson's efforts to create an innovative infrastructure are emblematic of great managers like Alfred P. Sloan, Jr. of General Motors who was a master at building organizational structures designed for greater efficiency and effectiveness.

Preparing for the Jet Age

Despite United's impressive growth in the years following World War II, Patterson remained vigilant and prudent, using lessons from the past to help navigate his airline's future. As we saw earlier, Patterson's experience with the Boeing 247 taught him the dangers of becoming too enamored with advances in specific airplane technology. The advances, it turned out, were often fleeting. Patterson applied this lesson to the dawn of the jet age. When the British came out with the Comet in the late 1940s, Patterson sent his Chief Engineer Jack Herlihy to Britain to report back. Although Herlihy felt the Comet was too small and too fragile to use as a reliable airliner, he did believe that jets were in United's future. To better prepare for that transition, Herlihy set up a "Paper Jet" airline service in 1952; it was a task force that would simulate the day-to-day running of jet airliners and would calculate the resources necessary to support it.[86] Through this process, United "simulated coast-to-coast round trips daily for two years, [and] gained invaluable experience, with meteorologists and dispatchers preparing the same forecasts and computations they would for actual flights."[87]

Aerial view of five DC-8 jetliners docked around the concourse projecting from the United Air Lines' passenger terminal at the San Francisco International Airport, California, circa 1965. (Source: Hulton Archive/Getty Images).

After the task force had supplied detailed information concerning United's future jetliner needs, Patterson eventually ordered 30 Douglas DC-8s in 1955 with a sticker price of $175 million. Instead of flying into a panic to acquire the best technology available during the mid-1950s (such as the Lockheed Constellation) or rushing to acquire the Boeing 707 jet because it would be delivered six to nine months earlier in 1959 than the DC-8, Patterson took the long view. For instance, although TWA's Super Constellation 1049G (known as the "Super G") would certainly give that airline an edge over United when it came out in April 1955, Patterson reasoned, "[t]his speed advantage will last for two, possibly three, years. But when we get our jets, we'll cruise . . . 125 miles per hour faster than any turboprop. On long, cross-country hauls, the turboprop will be an obsolete airplane, and before it's half depreciated at that. We've got to look further ahead. We can't take the risk of obsolescence before we get started."[88]

Constantly looking five, ten, even fifteen years ahead, Patterson had a long-term managerial perspective that enabled United to maintain its strength in the airline industry without incurring undue risk. Although not afraid to try new things, the stability of his company was first and foremost in Patterson's mind. Unlike many of the pilots who populated the industry, Patterson was not blinded by speed. Slow but steady, United trailed American Airlines for more than two decades after ceding their #1 position in the airline business in 1937.

With the acquisition of Capital Airlines in 1961, United Air Lines became the largest airline in the United States.

Capital Airlines

The acquisition of Capital Airlines was not without significant challenges. With 7,000 employees and a virtual hodge-podge of routes serving primarily the east coast of the United States, Capital was the fifth largest airline in the country and was on the verge of bankruptcy when it sought to be acquired by United. Patterson initially declined to acquire Capital Airlines, but rethought the decision when he secured the CAB's two-part agreement to complete the acquisition review process in six months and to refrain from selling off or awarding any of Capital's routes to other airlines. The acquisition was very difficult and caused "much temporary damage to United's reputation for on time performance and quality service."[89]

Patterson anticipated the difficulty and had created an integration team even before the merger to swiftly address both operational and cultural issues. Reflecting on the decision, Patterson commented:

> We took over Capital because I could see that if they went into bankruptcy, it would affect the jetliner financing of all airlines. Also, the jobs of 7,000 Capital employees were in jeopardy, as were the investments of 14,000 stockholders. They were the victims of management's mismanagement. Some of our rivals thought that we had bought a corporate corpse, but we worked hard to bring it back to health. We made sure that the former Capital people who joined the United force were not second-class employees in any way. They all got more money and better security and they have become some of United's best assets. We take more pride in that than in being the largest airline.[90]

After the Capital merger had settled and proved to be a success, Patterson made plans to leave the airline. He stepped into the chairman's role in 1963 and formally retired from all board activities in 1966. Reflecting on his success in the middle 1960s, Patterson discussed his approach working in the airline industry: "You have an idea? Hang onto it! . . . My heart still leaps when I see a tiny two-seater plane soaring gracefully through the sky. Our great airliners awe me. Yet I know they were not produced in a day or a decade. It may take years to put your idea into action. But if it has real worth, time will prove it and you will have something that will endure."[91]

PART III

The Leaders

To remain competitive and relevant, companies with dominant business models must adapt them to the changing contextual landscape. Managers who are successful for long periods of time are very innovative in their efforts to adapt. However, the innovations pursued by managers are often guided within the parameters of a defined business model. Managers either push the entire efficiency frontier onward or choose to differentiate some dimension of their company's product or service to remain relevant. Both approaches are designed to strengthen and reinforce a dominant business model within an industry; they typically are not designed to fundamentally transform the business. As managers work to reinforce their dominant positions, they may miss significant changes in the competitive landscape or they may be unprepared to handle fundamental shifts (like the move from regulation to deregulation) within an industry.[1] For some, they may be utterly incapable of making the shift. Their blindness can lead to opportunities for others to supplant their success or to build viable niches that have the potential to redefine the parameters of a new dominant business model. Failing to maintain a broad perspective on the environment can lead to a gradual erosion of power.

When the dominant business model within an industry begins to show cracks, managers often try to fill the cracks in the existing structure instead of looking to rebuild it or search for new options. Managers who have shepherded their companies through the growth and maturity phase of an industry often have a difficult time making the transition to a new model. Their very success in building and reinforcing their company's dominance within an industry often prevents them from engaging in activities that inevitably alter the foundation of that business model.

Success in an industry that is undergoing a decline or a seismic shift in the competitive framework requires the skills of the third archetype—the *leader*. Successful leaders seek to define a new business model within the midst of upheaval (*new business model leaders*) or look for ways of reestablishing a company's formerly dominant position (*turnaround leaders*). Through streamlining and reengineering, turnaround leaders make the often very tough choices needed to reframe the parameters of success. This is the key differentiation between managers and leaders. Managers push and extend known parameters of success whereas leaders reframe these parameters. The declining phase in an industry is usually characterized by a mad scramble to reduce costs and to build competitive scale through mergers that consolidate power. These efforts can often revitalize key companies, and in so doing, they can help to extend the competitive lifecycle of an industry. Alternatively, new business models emerge that can completely shift the industry lifecycle.

The personal computer industry has experienced this phenomenon, though its lifecycle has changed more rapidly than most. The industry began with several players in the 1970s, such as Apple, Tandy, Commodore, and so on scrambling to define a dominant position, each pursuing a very different and idiosyncratic approach that reflected the personality of their entrepreneurs. As Apple seemed to be gaining the dominant position in the marketplace, a number of competitors jumped into the fray. When IBM entered the market in 1980 with the aid of Microsoft's operating system, it established the de facto dominant model and supplanted Apple. IBM's approach was cloned by many others and the "IBM model" became the dominant player in the market. When Dell and Gateway introduced mass customization of personal computers in the late 1980s, they significantly realigned the economics of the personal computer market, and in the process, the industry lifecycle was regenerated. This process has continued to repeat itself with the introduction of better and faster microchips, hand-held devices, and other product extensions. Although many lifecycles as seen in the history of the personal computer industry are regenerated from within, others are modified by critical external forces. That was the case in the third phase of the evolution of the airline industry in the United States.

The third phase in the evolution of the airline industry was acutely punctuated by the move from regulation to deregulation though the competitive landscape had changed considerably in the decade before the formal passage of the Airline Deregulation Act in 1978. The increase in passenger travel in the 1960s sparked a series of investments in larger and more powerful aircraft. As he had done so many times before, Juan Trippe of Pan American was at the forefront of the investments in the latest jumbo jets, namely the Boeing 747s. To remain competitive, many of the major U.S. carriers followed Trippe's lead. As we saw in chapter 4, the complexity of the design of the jumbo jets significantly increased the development timeframe, and when the jets were finally available in the early 1970s, the competitive landscape of the airline industry had markedly changed. The country was in a midst of a recession and the oil crisis of 1973 deeply impacted the economics of operating such large aircraft. In fact, during the decade before deregulation, "fuel costs rose 222% (to 20% of operating expenses [up from 10%]); inflation boosted labor costs (to 45% of operating expenses); and the stagnation of GNP curtailed demand growth (to 4% from 18% per annum)."[2] In summary, the predictability of growth and the sustainability of the existing business model in the airline industry could no longer be assured.

The dramatic changes in the competitive landscape also coincided with a sweeping shift in the leadership of the major airlines. In a two year period between 1966 and 1968, Juan Trippe of Pan American, C. R. Smith of American, and Pat Patterson of United stepped down as heads of their airlines, and C. E. Woolman died while serving as Chairman of Delta. These four individuals led their respective firms for an average of 36 years—essentially throughout the regulatory and steady growth phase of the airline industry. Since 1968, Delta has had seven CEOs, American has had five CEOs, and United has had eight CEOs. Interestingly, American Airlines, which has had the fewest CEOs in this timeframe, has fared the best.[3] In many ways, the strong personalities of the initial leaders of these airlines and the tight control they maintained over their operations created a void in the succession planning process. All of their immediate successors struggled to sustain a viable business in the changing competitive landscape. Deregulation further changed the nature of competition and the leaders who ran these behemoths were ill-prepared for its impact.

The formal move toward deregulation was precipitated by a 1975 senate investigation of the airline industry led by Senator Edward Kennedy of Massachusetts. The investigation exposed the inefficiency of the major U.S. carriers compared to state regulated airlines that were very successful in California (Pacific Southwest Airlines) and Texas (Southwest Airlines). The major carriers or trunks (American, United, TWA, and Eastern—known as the

Four pioneers in the field of aviation get together in October 1968 at the Wings Club Annual Dinner. From left are: William "Pat" Patterson, retired president of United Air Lines; C. R. Smith, former chairman and chief executive officer of American Airlines; Captain Eddie Rickenbacker, former president of Eastern Airlines; and Juan T. Trippe, retired as chairman and chief executive of Pan American. (Source: Bettmann/CORBIS).

Big Four) had been lulled into a sense of complacency during their 40 year reign of the airline industry. During regulation, the CAB declined all requests to establish any new large-scale carriers; there were 16 established trunk carriers in 1938 and 9 by 1978, but throughout this period, the Big Four consistently controlled more than 70 percent of total U.S. market share.[4] Instead of creating direct competition for the major trunk lines, the CAB chose to award routes to several regional players like Delta, Continental, and Braniff that could provide service in areas that were underserved by the major lines or could provide a "feeder service" to the Big Four.[5] It was within this environment that a new breed of carriers emerged—the intrastate, local airlines.

Unlike the major or regional carriers, the intrastate lines were not regulated by the CAB. Senator Kennedy's investigation highlighted the fact that the intrastate carriers "attained higher levels of capacity utilization and superior financial performance with frequent point-to-point service at markedly lower fares . . . [T]he intrastate carriers believed that the market for air travel was much larger than the trunks or CAB imagined, and that the high price of tickets [supported by the regulated industry] stopped many Americans from flying."[6] The most successful of the intrastate carriers has been Southwest Airlines under the leadership of Herb Kelleher. The story of Southwest and Kelleher will be told in chapter 7. In many ways, Southwest has been able to define a new business model for success. Though many have tried to copy it (as we will see in chapter 8), few have been able to recreate the Southwest model.

In contrast to the cost efficiency of Southwest, the major carriers, under regulation, had little incentive to lower their cost structures. The carriers simply passed on any additional costs to consumers. Strikingly, "between 1968 and 1972 [the] average pay for airline employees rose by 51 percent at a time when average pay for all U.S. workers rose by only 28 percent."[7] The investigation of the CAB and the airline industry that showcased the inefficiency of the regulated business model also unleashed a torrent of consumer complaints about the artificially high cost of air travel. The formal report of the committee noted that "deregulation would allow pricing flexibility, which would stimulate new and innovative offerings; allow passengers the range of price and service options dictated by consumer demand; enhance carrier productivity and efficiency; and increase industry health."[8] Between the onset of the investigation in 1975 and the formal congressional votes on deregulation, there was an avalanche of antiregulation sentiment. The Airline Deregulation Act was overwhelmingly endorsed by the Senate in an 83 to 9 vote and by the House of Representatives in an equally lopsided vote of 363 to 8.[9]

The impact of deregulation on the formerly dominant carriers was devastating. Within two years of the passage of the deregulation act, the industry was in a tumult: there were 22 new low-cost carriers in the market; the total number of U.S. carriers had increased from 34 to 72; and all but 2 of the major carriers had lost money.[10] Despite all the upheaval in the airline industry and despite the number of new entrants in the market, eight major airlines still controlled 91 percent of all U.S. traffic.[11] In addition, the first decade after deregulation witnessed the birth and death of more than 150 carriers, several bankruptcies, and more than 50 mergers and acquisitions.[12] While Southwest was defining a new model for success, many traditional airlines struggled to survive through cost cutting of expenses and streamlining of services. In addition, traditional carriers sought to gain greater scale through targeted acquisitions and/or the creation of airlines within an airline (creating Southwest-like entities within their overall organizational structure). In chapter 8, we will examine one airline in particular that tried to do all these things—Continental Airlines. Continental under the leadership of Gordon Bethune in the 1990s epitomizes the archetypical turnaround leader story. The success of Continental in the aftermath of deregulation will be contrasted to the demise of Pan American at the end of the twentieth century.

Even the best efforts of leaders like Gordon Bethune, though, were not enough to recover from the terrorist attacks on September 11, 2001. The attacks obliterated the competitive landscape of the airline industry as well as many other industries in the travel and hospitality sector. The impact of this event combined with the resultant wars in Afghanistan and Iraq and the subsequent oil crisis have once again changed the economic structure of the industry. All carriers with the exception of Southwest Airlines have struggled to survive creating another wave of bankruptcies and consolidations. Since deregulation, profits for most airlines have been elusive. In the aftermath of 9/11, profits have been virtually nonexistent. The threat of future airline consolidations has the potential to lessen or even eliminate low cost air travel for the common person, which, ironically, has been a hallmark and justification of the deregulated industry. The airline industry once again stands on a precipice of monumental change. Will Southwest Airlines remain a competitive anomaly or will it become the new dominant business model? Will the former traditional carriers devise a new dominant business model? Will government step back in to reregulate the industry? If the role that leaders have played in the past in shaping the evolution of an industry is any indication, emerging leaders will undoubtedly play a key role in driving a new transformation process.

CHAPTER 7

Herb Kelleher at Southwest Airlines

Wearing satin shorts and a bathrobe, Herb Kelleher prepared to enter Dallas's seedy Sportatorium during a winter morning in 1992. Kelleher's "athletic" look, topped off with a sweatband pushing back his hair, was deliberately undercut by a cigarette dangling defiantly from his lips. Surrounded by an entourage rented from the world of professional wrestling, the 63-year-old CEO of Southwest Airlines had come to arm wrestle his company's latest nemesis: Kurt Herwald, the 38-year-old weightlifting CEO of Stevens Aviation, an aircraft maintenance company based in South Carolina. Herwald had recently taken offense when he learned of Southwest's use of the slogan "Plane Smart." It seems that Stevens Aviation had already been using "Just Plane Smart" as their slogan for the past couple of years. But instead of calling on his lawyers to get things straightened out, Herwald phoned Kelleher directly and challenged him to fight for the rights to the slogan, *mano a mano.* Seizing this as a chance for great publicity, Kelleher hyped the event as "Mallice in Dallas," gave 700 Southwest employees the morning off and brought them to the event to cheer him on. The arena soon echoed with the chant of "Herb! Herb! Herb! Herb!"

After staging an entrance that Gorgeous George would have been proud of, Kelleher got to business. Although he strained with all his might against his adversary's obvious physical advantages, Kelleher lost the match. "If it hadn't been for my hairline wrist fracture, my cold and my athlete's feet, I would have won," Kelleher protested. Gracious in victory, Herwald allowed Kelleher to keep using the slogan anyway. More important than winning the rights to the slogan, both men succeeded in obtaining great publicity for their companies. In the case of Southwest, Kelleher showed the public yet again that the nation's seventh largest airline had a lovable, zany sense of crowd-pleasing humor that came straight from the top.[1]

Although Kelleher's personality and fun-loving style certainly helped him to earn public relations points, one would be mistaken to assume that he was simply an off-the-wall businessman with a lucky streak. Spontaneous and unpredictable

Southwest Airlines CEO Herb Kelleher shows off his lighter side in this portrait, April 19, 1990 in Texas. (Source: Getty Images).

in public, Kelleher the CEO was as regimented and determined as an army general. Although Kelleher's antics brought the spotlight to his company, behind the scenes Kelleher was deadly serious about success. From his point of view, the airline business was "the closest thing to war in peacetime."[2] He welcomed the stresses the job could bring, explaining: "Life to me is a competition, and you distinguish yourself by succeeding in the competition."[3] At Southwest, Kelleher was legendary for his nonstop devotion to the airline, playing and working hard sometimes 16 hours a day for weeks on end.[4] Even with this frenzied schedule, Kelleher claimed he could remain calm and coolheaded, thanks to what one journalist called his "existential detachment."[5] As Kelleher himself declared, "You shouldn't get too heady about anything, because the greatest thing you do is not big in the universe. It's not saying it doesn't matter. It matters all the more. You're fighting against nothingness. But you don't give up. Therein lies the heroism."[6] Whatever his deepest motivations were, Kelleher certainly never gave up. His success lay not only in his indefatigable competitive spirit, but also in his ability to maintain a focused approach to running Southwest—an approach that transformed the tiny Texas commuter airline of the 1970s into the biggest success of the modern airline industry.

If one had to choose a single word to summarize the reasons behind Southwest's financial and business successes, it would have to be "discipline." Although Southwest's employees can be as zany as their CEO (in a flight to Austin during the 1988 Christmas season, "flight attendants were dressed as reindeer and elves, and the pilot sang Christmas carols while gently rocking the plane"), their commitment to the airline and maintaining its profitability provides evidence of a compelling company culture.[7] Maintaining longer hours than employees at other airlines, Southwest's highly productive workers during Kelleher's tenure were motivated by a vibrant working atmosphere, good wages, profit sharing, and the knowledge that no employees had ever been laid off at the company. This feeling of mutual respect and responsibility between management and employees inspired workers to take some extraordinary measures on behalf of their airline. For example, in the wake of the Gulf War and rising jet fuel prices, Southwest's Dallas employees initiated a "Fuel from the Heart" program in 1991 in which employees voluntarily incurred short-term payroll deductions to offset the firm's higher operations costs. Employees also more routinely accepted payroll deductions to help the families of fellow workers suffering from terminal illnesses.[8]

Southwest's operational approach to the airline industry was a curious niche model for many years, but it was a model that was well poised for the increasingly competitive and cutthroat context in the third phase of the industry's evolution. Founded 10 years before deregulation, Southwest's tightly integrated and aligned operational model established a new benchmark for success in the industry. The utter simplicity of Southwest's business model (no frills point-to-point air service) is one that many large, established carriers have struggled to emulate. The sheer complexity of the business models that worked well in the relatively stable confines of regulation were not at all suited to the new competitive landscape of the airline industry under deregulation.

Largely overlooked in the popular literature on Southwest, this culture combining wacky behavior and a deep commitment to the long-term health of Southwest and its employees was modeled after another airline with a very similar history. Pacific Southwest Airlines (PSA), which started in the late 1940s as a short-haul, intrastate airline based in San Diego, provided a model for Southwest. The former director of corporate communications for PSA maintains that Southwest personnel were trained by PSA in 1970; in addition, PSA allowed Southwest to use its own training manuals, "which gave the inexperienced airline 22 years of experience written down . . . as formulas for success."[9] In 1973, Southwest's head of marketing Jess Coker admitted as much when he stated that many of his airline's aggressive marketing campaigns were simply copied from PSA.[10]

PSA's Modest Beginnings

After having started a small company in 1946 that trained people how to fly private planes, Kenneth Friedkin (who would soon partner with his flying student J. Floyd Andrews) decided that he might make more money running an airline. Because he had very little capital, Friedkin decided to rent a DC-3 for $2,000 a month to serve a very specific niche market: San Diego to San Francisco. After its inaugural flight on May 6, 1949, PSA became a small intrastate airline, traveling mainly in the north-south corridor of California, which would eventually become the largest city-pair market for air passenger traffic in the nation during the 1960s.[11] To compete with the larger interstate companies such as United, PSA had to create a new kind of airline. J. Floyd Andrews recalled "that it was a tough nut to crack, but our fares were so ridiculously low that we attracted people who could not have otherwise afforded to fly. In the early days, we weren't competing with Western and United, we were competing with the train and bus."[12] PSA was so cheap that it earned the nickname "Poor Sailor's Airline" because of the many men from San Diego's naval base who took advantage of the low airfares. PSA's novel approach of competing with ground transportation with very cheap and frequent short-haul flights became one of the defining characteristics of Southwest Airlines in the 1970s.

By 1962, a couple of years after PSA made a major investment in new turboprop planes to fly its select few routes (Hollywood/Burbank was added in 1954), the airline had secured a 49 percent share of this lucrative niche market with profits of $1.4 million. Now businessmen were filling the seats, and the poor sailor's airline was becoming a major engine of California's economic growth. Despite its success, PSA executives were reluctant to travel outside of California in order to avoid coming under the jurisdiction of the Civil Aeronautics Board (CAB), which would probably not allow PSA to continue its low fares.[13] Although United Air Lines fought back by increasing its own commuter service and succeeded in decreasing PSA's share of the San Diego to San Francisco route, the number of passengers flying that route continued to grow from 1.2 million in 1960 to 2.2 million in 1964.[14] When Andrews took over as CEO in 1965, he decided to focus the airline's marketing efforts on catering to the businessmen who had become increasingly enthusiastic about PSA's schedule and consistently low fares. Besides serving cocktails, Andrews hit on a way to make boarding faster by eliminating advanced seat reservations, allowing passengers to sit anywhere they could find an empty seat.[15]

During this year, the year before Rollin King first proposed the idea for a new commuter Texas airline to Herb Kelleher, Andrews came up with a brilliant marketing scheme. With the Beach Boys singing the praises of the "California Girl," PSA would bring sexy California women in their planes by dressing stewardesses in fetching outfits: first, "banana skin" tight dresses and then miniskirts. Not only were the stewardesses attractive, they were also quirky, friendly, and unconventional—a reflection of the fun family atmosphere that CEO Andrews and his predecessor Friedkin had cultivated at PSA. Stewardesses sometimes surprised unsuspecting passengers by hiding in the overhead luggage compartment. During the last flight of one stewardess, her two colleagues "gave every passenger a party hat and horn and asked the group to sign a greeting card." And upon taking off, passengers might have heard this safety tip: "Place the oxygen mask firmly over your nose and mouth making sure you don't mess up your hairdo."[16] For the quirky and ambitious people founding Southwest Airlines in the late 1960s, PSA certainly seemed to have offered a model worth following.

Southwest's Early Years

Of course, the context of Southwest's beginning was different. Unlike PSA, whose early history was barely noticed by its competitors, Southwest believed it had a chance to make a big impact on the Texas market because the competition in the 1970s, for the most part, had grown complacent under the protection of the CAB and was alienating its customers. As one commentator lamented in 1976, "Even the high-priced advertising campaigns of the giants—with their talk about 'friendly skies' and their airplanes painted by famous artists—cannot entirely assuage the irritation caused by high fares, bad schedules, delays, and lost baggage."[17] The PSA model was adopted by Southwest because it provided a brilliant intrastate response to the national airlines that flew through Texas as part of longer, delay-ridden, and overbooked routes. Indeed, by changing a few details, the early history of Southwest sounds like a reenactment of PSA's strategy for success: offering low fares for frequent point-to-point travel that would become profitable through drawing new passengers out of ground transportation and into airplanes. To make the product even more attractive, Southwest also offered offbeat service and great-looking stewardesses on their planes.[18]

Similar to the founders of PSA, Southwest's co-founder Rollin King was a pilot. He made his living as an entrepreneur in San Antonio and ran a modest commuter airline. After studying PSA's success, King convinced his friend Kelleher, who worked as a lawyer in San Antonio, that a commuter-style airline with destinations to some of America's fastest-growing cities—Dallas, San Antonio, and Houston—might be as successful as PSA. In bringing along Kelleher to join his venture, King gave Southwest a huge, long-term advantage that PSA never enjoyed: the acumen of a hard-working and brilliant lawyer. Kelleher and King both knew that the airline would be vulnerable to the political and legal maneuverings of the two existing interstate airlines that already served the cities in that area, Braniff and Texas International. And because all U.S. airlines had been under the tight regulatory control of the CAB since 1938, they had become expert in wielding one of the few effective weapons available to them in their tussles with competitors: litigation.

But Braniff and Texas International would have to take on Southwest in new territory. Instead of the familiar corridors of the CAB in Washington, D.C., these two established airlines would have to meet the Southwest challenge at the Texas Aeronautics Commission (TAC), the regulatory authority in charge of *intrastate* airline traffic. Here, the playing field was leveled considerably, because Southwest's Kelleher was the ultimate Texas political insider. Although he was born in New Jersey and educated in the Northeast, Kelleher was first drawn to Texas during his undergraduate days at Wesleyan University; it was at this time that he met his future wife Joan Negly, who was attending Connecticut College. Joan was a native from San Antonio and part of a "wealthy

South Texas ranching family."[19] They married soon after graduation and moved to New York where Herb graduated with a law degree from New York University in 1956. After working as a lawyer for a few years in New Jersey, Kelleher decided to heed his wife's request and agreed to move the family to Texas. "He's never led a very structured life," commented his wife in 1985. "I think that's what he loves about Texas."[20] In San Antonio, Kelleher soon established a reputation not only as a good lawyer, but also as an astute political strategist, working to elect Senator Lloyd Bentsen as well as Governor John Connally. "Behind the scenes," commented San Antonio's city attorney in the 1970s, "he was probably one of the most powerful people in San Antonio."[21]

The nascent Southwest Airlines would need all the powerful friends it could get, and Kelleher had them.[22] After incorporating Southwest Airlines in March 1967, the founders of the airline worked hard to raise $543,000 for the battles ahead. "We initially figured we needed around $250,000," Kelleher later recalled, "but we doubled it because I was aware there was going to be a fight and it was going to be a prolonged fight."[23] The first skirmish came early in 1968 when Southwest appeared at the TAC to apply for what was called "a certificate of public convenience and necessity." The hearing drew a crowd. Lawyers from Braniff, Texas International, and Continental all argued that Southwest's application should be denied because there was just no need for a new airline. The six member panel of the TAC—most of whom were appointed by Kelleher's political ally Governor Connally—unanimously dismissed the complaints and gave Southwest permission to do business in Texas on February 20, 1968.[24]

The next day, the same three airlines that had lost at the TAC persuaded a state court to issue a temporary injunction against the commission's decision to allow Southwest to do business. Two years and many court appeals later, Southwest's competitors were denied an appeal to the U.S. Supreme Court. By that time, Southwest had been in existence for three years and still did not have an airplane. For those three years, Kelleher had been a one-man legal army, outnumbered by the big airlines' bevy of corporate lawyers; he had even been willing to work for no pay when the board of Southwest was willing to throw in the towel.[25] Although these challenges to Southwest had been a burden, Kelleher now believes they were crucial in his apprenticeship for the job of CEO: "Southwest Airlines would not be in existence today," he asserted, "had not the other carriers been so rotten, trying to sabotage us getting into business, and then trying to put us out of business once we got started. They made me angry. That's why Southwest is still alive. I'm not going to get beaten, and I'm not going to let anyone take advantage. They were too stupid to realize the psychology of the situation, so they just kept plowing ahead."[26] Here we see how much effort these leaders had to muster to fight the dominant model and how much resistance they encountered. In this regard, leaders are similar to entrepreneurs. Classic entrepreneurs engage in one form of creative destruction; they take on existing ways of doing things and create new industries (e.g., transportation by air instead of on the ground or by sea). Leaders engage in a different, but no less important, form of creative destruction. They take on an existing industry and reinvent it by introducing a fundamentally new business model that challenges the existing one. This act can generate as much resistance as the act of getting the market to accept something new.

Southwest's original route plan involved service between Dallas, San Antonio, and Houston, and its inaugural flight took off on June 18, 1971. King and Kelleher recruited Lamar Muse, retired from Universal Airlines, to run the company. Kelleher noted: "The directors hired Lamar and he was just perfect for getting it started. He was exactly what we needed. He was tough and iconoclastic in his thinking."[27] Muse helped to secure the initial financing to purchase aircraft for the airline and made the decision to secure the Boeing 737 that was offered at a discount from Boeing due to overproduction.[28] This is not uncommon. We often see the excesses of the dominant group open up opportunities for new models. For example, U.S. emphasis on steel from blast furnaces opened up the production of lower cost steel from scrap metal using a new production technology.

Muse helped to solidify the company's financial position when he oversaw the issuance of the company's first public stock on June 8, 1971 (10 days before Southwest's inaugural flight). Though the company struggled for the next few years, it gained considerable traction when it began to provide service to Hobby Airport in downtown Houston. Hobby had been essentially abandoned by the major carriers when the larger and more modern Intercontinental Airport was opened on the outskirts of Houston. Unimpressed by the location of the Intercontinental Airport and the lackluster service of the major carriers, business professionals flocked to Southwest's inexpensive and extremely convenient service to Hobby Airport.[29] With this move, Southwest adopted a strategy of servicing mostly second-tier airports.

Response to Local and National Regulation and Deregulation

With the inauguration of deregulation in 1978, both PSA and Southwest Airlines approached this new era in commercial airline history as an opportunity to expand their similar business models. PSA and Southwest both showed a profit in the first couple of years immediately following deregulation. Southwest earned a net profit of approximately $40 million from $300 million in total revenues in 1980 while PSA earned $20 million based on $350 million in revenues. After this initial period, the two airlines' earnings reports soon parted ways completely. Southwest continued to prosper, earning approximately $70 million in annual profits into 1986 while PSA fell from grace and permanently into the red.[30] Soon afterward, PSA itself would disappear and would be acquired by USAir in 1987.[31]

Who or what was to blame for PSA's demise? Both PSA executives and academics who have thought about the matter point to one major problem: the overweening regulatory limitations on airfares imposed on PSA by the California Public Utilities Commission (CPUC) during the 1970s. Although PSA was not a utility, the CPUC treated it as such. PSA's Andrews described the problem this way: "They [the CPUC] did not have a rate-making body that understood airline economics. We'd file for a rate increase and when it wasn't forthcoming, things would get worse and we'd have to file for a bigger one."[32] PSA officials protested that these increases were economically necessary and did not portend a future of higher airfares. "We're not trying to jack up the rates and have an instant goldmine," said a PSA lawyer during a dispute with the CPUC in 1978. "It would ruin the whole image. We pioneered low fares in California before we were closely regulated."[33] One expert on transportation describes the effect of the ceiling placed on PSA's fares by the CPUC: "PSA . . . was forced by the Public Utilities Commission to maintain its low fare structure prior to deregulation but needed to raise fares due to costs that were higher than the fares could sustain."[34] These assessments of the regulatory straightjacket placed on PSA seem to assume that the airline and its officials were essentially victims of unwise regulators; furthermore, these criticisms of the CPUC imply that PSA could do nothing but obey these governmental dictates.

Kelleher's early trials by fire provide a salient difference between the forces that helped to shape the leadership of Southwest Airlines' and those that formed the leadership of PSA. In the days before the Boeing 737s—the plane Southwest has used almost exclusively for thirty years—the small size of the few airplanes available to PSA in the 1940s and 1950s ensured that it would not seem to pose a threat to established behemoths such as United Air Lines. For PSA's CEO Andrews, political connections and legal maneuvering had little to do with his airline's early successes. So when the CPUC began to assert itself by limiting PSA's ability to raise airfares, Andrews felt unprepared to resist these challenges to his company's autonomy and was largely unable to counter this reassertion of the state's regulatory power. PSA's easy ride through the realm of regulations between the 1940s and 1960s made the obstacles put up by the CPUC in the 1970s seem almost insurmountable. In contrast, Kelleher's adept maneuvering through the courts, Texas's regulatory agencies and even Congress shows that PSA could probably have chosen to fight harder or maneuver smarter around government interference to defend its interests.

Kelleher's fighting spirit did not fail him when an even bigger obstacle appeared during the early days of airline deregulation: the powerful political opposition of Jim Wright, a congressman from Fort Worth and the majority leader of the U.S. House of Representatives. Deregulation offered Southwest the opportunity to fly outside of Texas without being hampered by the oversight of the now-defanged CAB. Lamar Muse, Southwest's CEO in 1978, decided that the first out-of-state destination would be New Orleans. Unfortunately, Southwest's success for the previous several years had stirred up some resentment in Texas. Specifically, Southwest had maintained the base of its Dallas operations from the city's old and modest airport Love Field, even though many local businessmen from the Dallas area had wanted to require all airlines to operate from—and help to pay for—the new Dallas-Fort Worth Airport (DFW). When Southwest refused to move to DFW (based on the fact that Southwest had never been party to any agreement concerning DFW), officials from the two cities and the new airport took the airline to court in 1972. Although the plaintiffs were defeated in 1975, Jim Wright championed their cause by passing a law in Congress four years later prohibiting interstate travel to and from Love Field. Undeterred, Kelleher went to the U.S. Senate (Wright's persuasive presence barred any opportunities in the House) with expert lobbyists to plead his case. Kelleher also contacted a friend from his law school days at NYU: Bob Packwood, who had become a U.S. Senator from Oregon. After months of negotiations, Southwest finally received the right to do business from Love Field to the four states bordering Texas. According to Kelleher, this new arrangement "is an unjustified pain in the neck, but not every legislative pain in the neck amounts to a constitutional infringement."[35]

Kelleher's philosophical stance toward Southwest's compromise with Jim Wright and the Dallas-Fort Worth establishment demonstrates yet another of his strengths: knowing when to quit. Unlike Eastern Airlines' CEO Rickenbacker or United Air Lines' Patterson in the 1930s and 1940s, Kelleher did not allow his personal grievances or a desire to win at any cost get in the way of the profitability of his airline. In the case of Patterson, his allegiance to his former boss Philip G. Johnson (who was barred from acting as United's CEO by the federal government in 1934) superseded the needs of his company. Although Johnson continued a successful business career despite this setback, Patterson insisted on pressing a decade-long suit against the federal government to redeem Johnson's reputation; he persisted even after federal officials had made it quite clear that the suit was dooming United's applications for new routes with the CAB. In contrast, although Kelleher strongly believed that the restrictions on Love Field were "unjustified," they were unjust compromises that Southwest could live with.

PSA's Diversification versus Southwest's Focus

Southwest's secret to success was to stick to a winning formula. Kelleher's great talent lay in not being wooed by the lure of quick profits or large market share to abandon it.[36] In countless articles about Southwest's success, Kelleher, company officials, and journalists have all pretty much agreed on many of the reasons why the airline has done so well since deregulation. In 1997, the *New York Times* added to the chorus of praise:

> Everywhere, Kelleher has applied what worked in Texas. Instead of a hub and spoke system that conveniences airlines more than passengers, he opts for flights averaging less than 500 miles. He favors scruffy, older terminals accessible to downtowns, lightly trafficked satellite airports and cities slighted by other carriers . . .
>
> To make training employees and servicing planes faster and cheaper, [Southwest] only uses one type of jet, the Boeing 737, requiring only a single parts inventory. To reduce financial risk, it avoids much debt, resulting in the highest credit rating and lowest borrowing costs among major domestic airlines. To avoid straining its finances or lowering its hiring standards, it usually adds only one or two destinations a year.[37]

One might add that many of these strategies for success had *already* been put in place long before Kelleher had taken formal control of the airline. For instance, Southwest's first CEO Muse decided upon the Boeing aircraft after seeing how PSA made use of the 737.[38] Using a single aircraft as a sole standard in an airline's fleet was ironically what C. R. Smith did when he introduced the DC-3 for American Airlines in the mid-1930s that significantly reduced the company's operating and maintenance expenses. As American Airlines served a broader network of routes, the company moved away from a standard aircraft and in so doing, substantially increased their operational expenses and the overall complexity of managing a diverse fleet. By 2005, American Airlines had a total of 699 aircraft comprised of 8 different models. In contrast, Southwest's entire fleet of 445 aircraft was comprised of Boeing 737s.[39]

Alas, for PSA, simplicity was not its forte, at least not by the 1970s. Even before its problems culminated with the CPUC, Andrews allowed his impatience for profits to divert his attention toward diversifying PSA, which resulted in largely unproductive ventures. In 1967, PSA opened a flying school. Soon afterward, the airline founded an airplane leasing company in response to its inability to sell some Boeing 727s it wanted to dispose of. Going further away from the business of running an airline, PSA then tried to create a vacation industry empire of sorts. So much was going on that a new holding company called PSA, Inc. was created to organize these activities. Most of these ventures quickly lost money. In the end, Andrews's impatience moved him to grasp at financial straws instead of honing his vision for success in his realm of expertise, the airline industry.

In contrast, Southwest delayed or abandoned plans for expanding existing routes or moving into a new market if either expansion required a major modification to its usual business procedures. In fact, Southwest's first CEO may have left because he was ready to betray the airline's prudent approach to expansion.[40] Newspaper reports about Muse's departure from Southwest in 1978 focus on the dispute between Muse and many other Southwest executives concerning Southwest's plans to serve Midway Airport in Chicago. Muse was impatient to serve Chicago, but he was rebuffed by other executives who feared that entering that market would risk placing all of Southwest under the regulatory authority of the CAB.[41] Under Kelleher, Southwest was content to do without the Chicago market until 1985 when it introduced service between Chicago and St. Louis.

Kelleher—The Founder as CEO

Kelleher's path to the position of CEO was a bit circuitous, making it difficult to say when exactly he took full control of the airline. In August 1978, Muse's successor Howard Putnam was brought over from United Air Lines, where he had been group vice president of marketing services.[42] To smooth his transition to Southwest, Putnam shared responsibilities—Kelleher became chairman of the board while Putman assumed the role of CEO and president of the airline.[43] In three short years, Putnam was lured away again, this time to rescue the ailing Braniff Airlines.[44] During this time, Kelleher familiarized himself more with the details of the daily operations of the airline—a process that turned out to be his CEO training.

The wide-open playing field of deregulation prompted many airline CEOs to take very bold moves to secure a dominant position within a vastly different competitive landscape. Most analysts of the airlines believed that "meals, pre-assigned seats . . . and a hub and spoke route system were critical to the success of an airline going forward."[45] The hub and spoke system impacted airline passengers the most. With airfares no longer maintained at artificially high rates by the CAB, the hub and spoke system promised to reduce labor costs and improve the bottom line. For many of the major airlines, routes would be channeled through a few central "hubs" where long-haul passengers from a variety of destinations (the spokes) would be directed to change airplanes to continue their journeys. By concentrating much of their operations in hubs, airlines hoped to decrease labor costs and keep passengers from transferring their business to other airlines in their staggered flights across the country.[46] Some hubs became sources of competitive advantage by permitting certain carriers to dominate air passenger traffic in particular parts of the country. In 1992, for example, Delta Air Lines "leased sixty-six of Hartsfield [Airport's] 146 gates, accounting for 89 percent of passenger traffic at the airport."[47] By 1997, Delta became the first airline to board more than 2 million passengers in one city (Atlanta) in a month.[48]

Although airlines were happy with the presumed financial advantages of the hub and spoke model, passengers found that scurrying through airports to catch their connections hardly marked an improvement in airline service. Moreover, the efficiency of the hub and spoke model for passengers depended on two or three airplanes flying routes spanning the nation to arrive and depart *on time.* The model was prone to a myriad of potential delays caused by bad weather, mechanical problems, or human error. Indeed, with the airlines' rush to find efficiencies to compensate for falling fares, most industry experts believed, even long after deregulation, that a decrease in the already middling quality of airline service had been *inevitable.* As one journalist noted, many airlines cut costs in passenger service wherever they could right after deregulation: "Many furloughed flight attendants and reservations agents, reducing in-flight service and lengthening the time needed to book flights. Some put more seats on the planes, reducing legroom, in bids to squeeze as much revenue as possible from each flight."[49] Richard Ferris, one of Patterson's CEO successors at United, once responded to a question from Boeing on how best to configure the seating arrangements for his passengers by saying: "Don't bug me about double-aisle or seven abreast [seating]. I want the most efficient airplane. Just guarantee the seat-mile performance. As for customer preference, I couldn't care less."[50]

From the late 1970s into the 1980s, Kelleher was almost never tempted to stray from Southwest's productive niche in the industry. He ignored the analysts because he believed that he had found a new and sustainable market for convenience and affordability. Having already adopted and enhanced a formula of outstanding (albeit no-frills) service for passengers along with healthy profits for Southwest Airlines, Kelleher felt no pressure to change the company's basic approach, except, perhaps, to be even more vigilant about keeping costs low and efficiency high.

More importantly, Southwest chose to grow slowly after deregulation, even though the whole of the United States and perhaps even much of the world lay open as possible markets for Southwest to exploit. In Southwest's annual report for 1979, Kelleher laid out the path for growth that he would follow almost religiously until his retirement:

> We are pleased that during the first two months of 1979 our revenue passenger miles increased 45% over the comparable 1978 period, while our load factor went from 61.33% to 65.95%. Although this demonstrates the significant growth potential of our present routes, we continue to study very carefully additional short-haul, out-of-state possibilities affording us the same kind of passenger volume and return that we have experienced to date. Under the Deregulation Act of 1978, new opportunities for new routes will continue to present themselves and give us the potential for *strong future growth on a planned and controlled basis* [emphasis added].[51]

True to his word, Kelleher permitted Southwest to expand past its Texas roots slowly but surely. After adding New Orleans as its first interstate route in 1979, Southwest waited until 1982 to make significant, yet relatively modest, strides outside Texas. In that year, the airline added five new cities outside of Texas, including Phoenix and San Diego.[52] A few years later, when the opportunity arose to strengthen its foothold in Chicago, Southwest pounced. The other airline that served Midway Airport, Midway Airlines, had been doing poorly for years and went bankrupt on November 13, 1990. The next morning Southwest executives rushed to seize this new opportunity; they arrived in Chicago at 9:00 a.m. and signed a deal with the city by 2:30 p.m. to take Midway's place that same day.[53]

Other airlines succumbed to the temptation of expanding too quickly. Southwest's regional competitor, Braniff Airlines, bid for a whopping 626 new routes following deregulation, "far more than any other carrier."[54] Just a few years later, Braniff was pulled down by the massive debts incurred to reach for its starry-eyed ambitions and became one of the first of the major airlines to declare bankruptcy from which it never recovered.[55]

Besides fiscal prudence, Kelleher felt that there was another important constraint on growing Southwest too quickly: maintaining the company culture. Kelleher felt that the extended family feel of Southwest could not be maintained if too many new workers were integrated at one time, or if many workers were ever laid off. Looking back at Southwest's long history of good relationships with its employees, Kelleher remarked:

> The thing that would disturb me most to see after I'm no longer CEO is layoffs at Southwest. Nothing kills your company's culture like layoffs. Nobody has ever been furloughed here, and that is unprecedented in the airline industry. It's been a huge strength of ours. It's certainly helped us negotiate our union contracts.... We could have furloughed at various times and been more profitable, but I always thought that was shortsighted. You want to show your people that you value them and you're not going to hurt them just to get a little more money in the short term.[56]

This attention to the company culture reaped great benefits for Southwest.[57] Southwest employees' dedication combined with Southwest's exclusive use of the Boeing 737 were the key elements behind stunning aircraft turnaround times (the amount of time an aircraft is on the ground between landing and its next take-off) that were consistently far below the industry average, thus keeping each of their planes in the air longer and substantially decreasing the amount of aircraft needed to transport Southwest's passengers. Kevin and Jackie Freiburg, the authors of *Nuts! Southwest Airlines' Crazy Recipe for Business and Personal Success*, explain the benefits of this for passengers: "The fifteen- and twenty-minute turns let Southwest use about thirty-five fewer aircraft than an airline with an industry-average turnaround time. With the cost of a new 737 at $28 million in 1995, it's not hard to figure out the savings: $1.3 billion in capital expenditures, which is, in turn, passed on to the customer in the form of lower fares and to the shareholder as profits."[58]

Great employee relations also staved off labor problems that weighed down other airlines. For example, after PSA struggled out of its mistakes of the 1970s and attempted to take advantage of the opportunities opened by deregulation, pilots mounted a strike in 1980 to protest the airline's plans to reduce the number of people working in the cockpit from three to two. According to one historian of PSA, it took five years to make up for the passenger losses incurred due to the strike, which shut down the airline for 52 days.

Southwest versus Traditional Carriers: Different Approaches to Expansion

Rejecting the fiscal prudence that Kelleher embraced, American Airlines' CEO Robert Crandall attempted to dominate the airline industry through a massive $20 billion dollar investment in new airplanes and new hubs, starting in 1983. Crandall's approach was in keeping with the former dominant business model that benefited from bold investments in new equipment. In many ways, Crandall's actions were a seemingly natural follow-up to the efforts pioneered by Trippe and Smith. Unfortunately for American Airlines, the context of the airline industry had changed dramatically.

Despite these changes, much of the industry—with Southwest as a notable exception—followed Crandall's lead; ten years later, one industry expert commented, "the strategy did deliver a big share of the market to a handful of carriers. But it produced not a penny of profits."[59] In the case of American Airlines, Crandall's long-term investments depended on low labor costs and high passenger traffic to translate into profits. The combination of labor's successful fight in 1991 against American Airlines' attempt to lower wages combined with the "plateau in passenger traffic that developed after 1987" doomed Crandall's plans.[60] In contrast, Kelleher liked to keep his investment in infrastructure, such as large hubs, as low as possible and favored instead to nurture Southwest's capacity to adjust quickly to changing circumstances. As Kelleher explained, "You become used to a life of quick change. This is an operating business. There's nothing passive about it when your principal assets are moving at 540 miles an hour."[61]

Crandall's investments depended on the ability to predict future business conditions years in advance, a strategy that Kelleher saw, frankly, as impossible to realize in the turbulent airline industry. Lean and quick, Southwest avoided the poison pill that American Airlines first took in 1983: overcapacity.

Although union demands for high wages contributed to the deep hole many airlines dug for themselves after deregulation, airline critic Stephen Solomon insists that the problems caused by industry leaders were just as harmful: "a fundamental problem that the industry has never solved [is] too many seats and too few passengers. Led by Mr. Crandall's American, the industry went on a decade-long binge of buying big jets, adding seats at a pace far exceeding demand."[62] The contrast between Kelleher's consistently upbeat, ready-for-anything demeanor in 1992 and Crandall's sad self-assessment in 1993 is striking. Reflecting on his company's poor performance in the early 1990s, Crandall concluded gloomily: "I haven't had any fun for the last three years. I work like hell and at the end of the year we have a big loss. That makes me a loser. Nobody likes being a loser."[63]

By the mid-1980s, many of the larger carriers started to imitate Southwest's trademark low fares. "They've all been to Southwest University," Kelleher chuckled ruefully, reflecting on big carriers such as American Airlines and Continental introducing lower fares with catchy names like "MaxSaver."[64] Although the major airlines' decision to feature low fares might have, according to Southwest's COO, caused Southwest to lose "some of our image as a

Southwest Airlines founder Herb Kelleher, left, and former American Airlines CEO Robert Crandall pose for a feature about local inventors who changed the way we live, 2005. (Source: Natalie Caudill/ Dallas Morning News/CORBIS).

low-fare leader," Kelleher was not too publicly worried about the new trend. "We don't compete with Eastern or Continental to a great extent," he remarked.[65] When Southwest did go head-to-head with smaller start-ups, such as it did with America West in the Phoenix market in the late 1980s, Kelleher's airline came out the victor. With its very low costs and high profits, Southwest could wait out a fare war with smaller airlines.[66] Moreover, Kelleher openly scoffed at America West's aggressive growth strategy, which many analysts cited as part of the reason the company filed for bankruptcy protection in June 1991.[67]

Although the arc of Kelleher's accomplishments in the 1980s was very impressive, it was not devoid of mistakes in judgment. Around the same time Southwest was winning fare wars with airlines like America West, Kelleher did something that was quite uncharacteristic: he began to move away from Southwest's formula for success in 1985 when he bought an airline that differed greatly from the Southwest model. That airline was "Muse Air," founded by the former Southwest CEO Muse in 1981.[68] Dubbed "Revenge Air" by industry insiders, "Muse Air" attempted to be everything Southwest was not—a classy, long-haul *and* short-haul airline, featuring frilly service and reserved seating on flights to Florida and California. Most of Muse's routes shadowed those of Southwest.[69]

Muse's attempt to get back at Southwest failed and his eponymous airline went into bankruptcy after a few short years. Although Kelleher told the press, of course, that buying Muse Air was a great opportunity for Southwest, its acquisition fit awkwardly in Southwest's history of growth and expansion. While the newly acquired airline (renamed TranStar in February 1986) contributed to Southwest's bottom line, it was treated as a separate entity. The personnel from the two airlines did not interact much and their pay scales were different, which was hardly typical of the warm, open, and egalitarian working conditions Southwest prided itself upon.[70] Most significantly, many of Muse Air's routes were longer than Southwest's, such as a direct flight from New Orleans to Miami. When Continental saw this development, it applied its superior resources to compete against Muse Air, and the airline was unable to keep pace.[71] Southwest had to pour some of its profits into Muse Air to shore up the struggling airline.[72] Luckily for Kelleher, his better judgment moved him to withdraw from this unfamiliar arena when it seemed that a fight with Continental was not worth the risks.[73] Kelleher liquidated the struggling airline in 1987.[74] Looking back on the acquisition, Kelleher took this lesson: "when something turns into a financial mistake, just stop it."[75]

Despite the ill-fated decision to acquire Muse Air, Southwest posted $778 million in total revenues in 1988, and seven years later, revenues more than tripled to $2.5 billion. As Southwest explained to its stockholders in its 1994 Annual Report, since "1990, the industry has been shrinking and we have been expanding. We are now carrying more than twice the number of passengers annually than in 1990, an annualized growth rate of 21 percent."[76] Kelleher's plans to expand his company somewhat more aggressively, though still prudently, starting in 1989 were paying off tremendously, and much of that expansion was facilitated by his shrewd decisions and attention to detail.[77] For instance, because most airlines struggled to cut costs from their prederegulation standards, rising prices of airplane essentials such as fuel were difficult for them to absorb. In contrast, Southwest's high profits, cash reserves, and already rock-bottom fares allowed them to raise their ticket prices slightly without deeply affecting their bottom line or market share.[78] Kelleher explained in 1991 that he "started cutting costs right after our all-time record second quarter in 1989 because I anticipated we were heading into a recession."[79] Although hindsight can be 20/20, events of 1990 seem to confirm Kelleher's prescience because Southwest was the only major airline to record a profit that year.[80] Keeping an eye on the economy and its impact on the company's profitability has been vital to Southwest's practice of hedging its fuel costs. By 2006, Southwest's fuel hedging strategy, which it began to aggressively pursue in the mid to late 1990s, has saved the company more than $2 billion.[81]

Kelleher also remained an expert at political and legal maneuverings to expand his airline and hamstring the competition. In 1991, Kelleher took the lead in Texas to work against the introduction of high-speed rail. He claimed that this new rail system could only exist with massive government subsidies. Although he may have had a point, Kelleher's passionate arguments against high-speed rail were strained, considering that the cars and buses he competed with also benefited from government subsidies in the form of federal funding for highways. The effort to bring in the French T.G.V. high-speed trains eventually collapsed in 1994 after Kelleher threatened the Texas legislature that he would move Southwest's headquarters away from Dallas if this new railroad were to be built. Although Kelleher might be seen as a generally fair and straight-talking CEO, when Southwest was deeply threatened, in this case by high-speed trains, Kelleher pulled out all the stops to protect his company.[82] In this sense, Kelleher was no different than the executives of the major airlines who had fought to block his entry into the Texas airspace. He defended his company against anyone who would threaten it, especially a new business model.

The Southwest Effect

By 1993, Southwest Airlines had become such a formidable player in the airline industry that the U.S. Department of Transportation actually came up with a name for its impact on any particular market it entered: the "Southwest Effect." When Southwest entered a market, fares decreased 65 percent, passenger traffic increased 500 percent, and its competitors were often forced to increase their flights into that market as well.[83] But 1994 would bring an unusual challenge to Southwest.

The United Shuttle was launched in 1994 in an effort by United Air Lines to win back the market share of intrastate travel in California that it was losing to Southwest. In that year, Southwest held approximately 50 percent of the so-called California Corridor market between Los Angeles and the San Francisco Bay area, which was flown by more than 12 million passengers in 1993, surpassing United Air Lines as the market share leader.[84] These routes were doubly important for United because many of them fed passengers from California into the carrier's long-haul routes.[85] Although Southwest was certainly the industry's juggernaut in profits, it was a relatively small airline in revenue terms compared to United that had total revenues of $15 billion, which seemed to afford the airline a cushion of cash to withstand a probable fare war with Southwest. United Shuttle tried to differentiate itself from Southwest by offering better service (accomplished mainly through featuring reserved seating) and giving passengers the opportunity to combine their short-haul and long-haul frequent flyer miles.[86] The Shuttle's main problem would lie in cutting its costs to the ambitious goal of 7.4 cents per passenger mile (United's normal passenger mile costs were 10.5 cents), just more than Southwest's level of 7.1 cents per passenger mile.[87]

Kelleher responded with a call to arms; in so doing, he showed brilliance in communicating his vision and the ability to galvanize his employees behind him. In November, he distributed a video to all Southwest employees in which he treated the introduction of the United Shuttle as a potentially fatal rival of Southwest Airlines: "The United Shuttle is like an intercontinental ballistic missile targeted directly and targeted precisely at Southwest Airlines and no other carrier in the world," Kelleher proclaimed. In addition to motivating his workforce, Kelleher fought back with adding even more flights to an already busy flight schedule in California and introduced occasional special rates that were even lower than Southwest's usual $69 fare for intrastate flights in California, such as one-way fares on Thanksgiving for less than $10.[88] Although United made some early gains, it could not match Southwest's lower costs per passenger mile nor Southwest's consistent service. By 1996, United could only push costs down to approximately 8 cents per passenger mile. Practically conceding defeat, the Shuttle retreated and only operated flights that connected to its San Francisco and Los Angeles hubs, where heavy traffic caused frequent delays.[89] By 2001, United called it quits and discontinued its Shuttle service.

The End of an Era?

Although Southwest's employees have long been encouraged to take their own initiative to find ways to make the airline run more efficiently, Kelleher was the inimitable heart and soul of the airline until he decided to leave his position as CEO at the age of 70 in 2001. Finding a successor was certainly a difficult task, and not one that Kelleher was eager to discuss in public.[90] In the end, Kelleher may have attempted to maintain control in spirit, if not completely in person. He chose a long-time labor negotiator at Southwest, James Parker, to take over as CEO; Parker was a lawyer and had come over to Southwest after having worked for Kelleher's law firm in San Antonio. In addition, Kelleher placed his very able and long-time assistant, Colleen Barrett, in the role of president and COO. Barrett, besides being an astute business person, would also nurture the family atmosphere Kelleher worked so hard to establish and maintain.[91] If continuity and stability were important for Kelleher's Southwest legacy, then one might conclude that Kelleher may have failed. Just three years after becoming CEO, Parker resigned, citing a desire to spend more time with his family. But Parker's departure had been preceded by some uncharacteristically contentious public disputes in labor negotiations with Southwest's flight attendants. Things were so bad that Kelleher, who had retained a role at Southwest as chairman of the board, had to step in to avert a strike.[92] Gary Kelly who joined the company in 1986 as its controller and who became its chief financial officer in 1989 was named as Parker's successor.

Before his formal retirement, Kelleher had also maintained that the culture of Southwest was so strong, so ingrained in the company that no one person, not even a CEO, could adversely affect the overall strong performance of the airline. Southwest, Kelleher said, "is an airline of collaborationists, collective leadership."[93] Since his retirement in 2001, the company has continued to achieve the strongest profit margins in the industry; it is usually the only profitable airline.[94] The business processes and procedures that PSA had originated and that Southwest perfected became the model for airline success in the post-regulation environment. Kelleher nurtured, cultivated, and protected this new model, and his successors have continued to build upon that success. As Kelleher leaves the board in 2008, the real test for the sustainability of the "Southwest model" will begin.

CHAPTER 8

Gordon Bethune's Revival of Continental Airlines

When Gordon Bethune took over Continental Airlines in October 1994, he and his close associate Greg Brenneman had little time to overhaul the airline.[1] Net income had plummeted from a loss of $113 million in 1993 to a staggering $613 million deficit in 1994.[2] Bankruptcy and liquidation were not just on Continental's horizon; they were pounding on the front door. For a week, Bethune and Brenneman spent the evenings together debating how exactly they would save their moribund airline. These "last suppers," as they ironically called them, turned out to provide the insight and energy that served to resurrect the airline. Brenneman describes the essentials of their approach: "Most companies that are in trouble... tend to develop a myopic focus on cost. They forget to ask simple questions like, Do we have a product people want to buy?... [T]hey forget to think about money in, or good old revenues."[3]

Although this business philosophy might seem far from rocket science, Bethune's eventual accomplishment was nothing short of astronomical. After a long series of mishaps beginning in the 1980s, Continental had drifted far away from being the proud and profitable airline that had been founded by one of the industry's fabled pioneers, Robert Six (see sidebar). By the 1990s, the popular press reported that Continental employees were so demoralized and ashamed of their company that many ripped their Continental badges right off their shirts when they left work.[4] As Karen Radabaugh, Continental's Airport Training Manager, said: "If someone asked me what my job was, I would say, 'I work for an airline.' I did not want to say 'Continental Airlines.' "[5]

In 1994, the flying public had already become accustomed to airline bankruptcies. Since Braniff's demise in the early 1980s, many household company names, including Juan Trippe's imperial Pan Am, disappeared off America's corporate radar screen. Bethune, unlike many other would-be leaders of the airline industry, was able to stop his airline's financial bleeding and turn around a company culture that had been degenerating for approximately 15 years. By 1996, Continental's net income had increased to a positive $319 million. In short, as described in the title to his popular book, Bethune was able to transform the company "from worst to first" in the airline industry. By comparing Continental's turnaround to Pan Am's fall, we can see clearly how two airlines confronted the challenges of airline deregulation.

Specifically, in the case of Continental in 1994, Bethune found his way around the conundrum that plagued his predecessors at Continental and his counterparts at Pan Am: in planning a corporate recovery, should a CEO favor cutting costs and service to improve the bottom line or should a CEO improve service (and probably increase costs) to draw in more customers? Bethune's genius was to refuse to see the demands of service and profits as a zero-sum game. In Bethune's calculus, one could improve both simultaneously.

Beginning of the End: Pan Am's Trippe and Continental's Lorenzo

Like Trippe, Frank Lorenzo (who acquired Continental in the early 1980s) possessed a great ego, a swashbuckling approach to expanding his business, and the desire to rule over an empire. Both men were also fascinated with planes from an early age and devoted themselves to the airline industry early in their business careers. But Lorenzo and Trippe rose to prominence in very different eras, which helps to explain some of their most pronounced differences. For Trippe, making money was a means rather than the goal of building Pan American World Airways. Trippe used his financial connections to bail him out of some of his most adventurous miscalculations in the 1920s and later employed his government connections to extend Pan Am's monopoly over international flights originating from the United States. But whatever the era or the circumstances, Pan Am was Trippe's chosen instrument of extending his personal power in the world. His desire to be master of the international skies manifested itself in Pan American's almost unbelievably extensive global route system by the end of his tenure in the late 1960s. The attitude at Pan Am was that profits would follow because they always had.

Lorenzo was not an aviation pioneer creating an international airline empire from the ground up; instead, he used aviation to construct a financial empire. An MBA graduate of Harvard Business School in 1964, Lorenzo came to prominence in the 1980s during the rise of financial wheeler-dealers, such as Henry Kravis and Michael Milken, who, respectively, came to be associated with leveraged buyouts (LBOs) and junk bonds. Deregulation in 1978 exposed the airlines—which had formerly been insulated from the designs of unpredictable financial wizards—to the possibility of being bought, traded, sold, and taken apart just like most other American companies.

Although Trippe and Lorenzo achieved great things, their legacies remain mixed because of a weakness they shared: an unrealistic or underdeveloped sense of how to sustain the success of their airline ventures. In addition, their egotistical personalities left considerable uncertainty and disarray among executives and staff. Neither truly groomed a successor from within. Trippe, as we have read earlier, kept most of the decision making at Pan Am an intensely personal and secret matter. Through a myriad of successors, Pan Am was never able to recapture its pioneer spirit or profitability. And unforeseen disasters such as the terrorist bombing of a Pan Am plane over Scotland in 1989 and the Persian Gulf War of 1991 easily undercut attempts to save Pan Am. The last plan, to convert Pan Am into a smaller airline serving mainly the Caribbean and Latin America, fell apart: Delta Air Lines refused to lend cash to Pan Am after acquiring its East Coast Shuttle and European routes in December 1991.[6] After 64 years of operation, Pan Am ceased to exist. Continental was destined for the same fate in the early 1990s, due in large part to the actions of Frank Lorenzo.

Lorenzo the "Magnificent"

Lorenzo, who became CEO of Texas International Airlines in 1972, responded immediately to airline deregulation in 1978 by hiring Kidder, Peabody and Smith, Barney to make independent assessments of how Texas International might take over Continental. By late 1979, Lorenzo had the results. Both investment-banking firms advised that a takeover could be financed with Continental's own assets as collateral.

Continental employees were well aware of Lorenzo's efforts to accumulate an increasingly larger percentage of Continental stock, and they were not surprised when in February 1981, Lorenzo tendered a hostile takeover bid, agreeing to pay $13 per share to stockholders not already part of the 48.5 percent he controlled. It was the latest move by the airline executive that James Cook of *Forbes* earlier called "Lorenzo the Presumptuous," for trying to buy much bigger airlines with a questionable debt structure.[7] Continental's founder, Robert Six, returned to the helm of the company to help thwart Lorenzo's takeover bid and salvage the company's reputation and heritage (see sidebar).[8]

Robert Six—Continental's Heritage

Six bought Varney Speed Lines, the precursor to Continental, in 1936. At the time, the company had one airplane and one primary route (Denver, Colorado to El Paso, Texas) for airmail and passengers. For the next 44 years, Six built the airline based on a strong attention to detail and a heightened sense of quality. He believed that a strong focus on the customer and a dedicated attention to quality would enable Continental to distinguish itself in a regulated market.

Throughout the 1930s and 1940s, Continental remained a regional carrier, known for excellent customer service. The airline grew dramatically in the 1950s and 1960s through small, targeted acquisitions and through the introduction of Six's innovative Gold Carpet Service Program for passengers en route from Chicago and Denver to Los Angeles. Service on Continental's Gold Jets was first-rate. A crew captain handled all ticketing and baggage functions on the plane itself, and each plane offered a concierge type service that included a special radio phone that passengers could use to set up hotel and rental car arrangements in either Chicago or Los Angeles. While on these planes, Continental offered customers "Country Club" quality cuisine with a choice of four entrees. In addition, there were two separate lounges on the DC-7Bs including a stag lounge just for male passengers. Six oversaw every aspect of the program. Continental's employees responded well to the airline's quality focus; they were proud to be associated with the company and worked hard to deliver on its business objectives.

Through a dedicated focus on the consumer and through his tight management of the company, Six built Continental from a small regional airline carrier to a dominant player with close to $1 billion in revenue by 1980.

In November 1982, over a year after attempting his hostile takeover bid, Lorenzo officially gained control of Continental, setting into motion a series of mergers in the next few years that had a significant effect on the company's future.[9] Chief among the legacies of the Lorenzo era were massive layoffs and bitter management-employee relations, engendering a bureaucratic culture where internal silos were built to fortify departments and employees against downsizing.[10] These attitudes became entrenched after Lorenzo's announcement on September 24, 1983 that Continental would rely on a recent change in federal bankruptcy laws to declare Continental bankrupt to sever all existing union contracts and renegotiate lower wage and compensation packages for Continental employees (even though Continental still had $25 million in cash reserves).

Lorenzo later explained that Continental's Chapter 11 bankruptcy filing in 1983 was in keeping with one of his basic goals for Continental. "What we are looking to do coming out of bankruptcy is to have a company that will survive and be stable through the 1990s, strategically sound and diversified."[11] Others claimed that Lorenzo was merely trying to break the power of the unions to renegotiate lower wage and compensation packages for Continental employees. Over a two day span, Lorenzo cut a staff of 12,000 people to 4,000, and then gradually replaced the 8,000 workers who were dismissed with nonunion workers at lower salaries and benefits. The impact on Continental's bottom line was sizeable; by the close of 1984, Continental achieved a record annual net profit of $50 million.[12]

These healthy bottom line figures masked a growing problem at Continental—sagging morale. Rather than work on the necessary synergy to combine Texas Air with Continental, Lorenzo's primary focus was expansion through more mergers and acquisitions. While Continental was still in bankruptcy, Lorenzo joined forces with Michael Milken of Drexel, Burnham Lambert to design the financial terms of Texas Air's takeover bids of TWA and Eastern with junk bonds. Although Lorenzo missed on TWA, he did secure Eastern in February 1986.

Later in that year, Continental emerged out of bankruptcy protection and Lorenzo continued to attempt more mergers. In February 1987, Continental acquired Frontier, People Express, and New York Air to become one of the largest U.S. airlines at that time. Don Burr, CEO of People Express when it was purchased by Texas Air, offered an important insight into Lorenzo after resigning from Texas Air later that year. "Frank is very simpleminded: you do whatever you have to do to make money. Any means to victory is okay. Frank is capable of any type of behavior to win."[13] Communication at the executive level became strained after four of the presidents Lorenzo hired—Thomas Plaskett, Martin Shugrue Jr., D. Joseph Corr, and Mickey Foret—each left the company after a year or less on the job.

The size of the airline did not offset the piles of debt that Lorenzo accumulated for Continental. In 1987 and 1988, Continental lost a combined $500 million and carried a debt of $2.3 billion. In response, Lorenzo asked his increasingly understaffed and demoralized employees to help him by cutting costs wherever and whenever possible. To help stem losses, Lorenzo and Continental set up an international partnership with Scandinavian Airline Systems, SAS. However, a key condition of the partnership arrangement with SAS was that Lorenzo would have to step down as CEO. He did so taking with him a multimillion dollar compensation package.[14]

Lorenzo was still on Continental's board of directors when his successor as CEO, Hollis Harris, replaced him in August 1990. Harris had occupied the COO position at Delta Air Lines, having risen from his first position at the airline as a transportation agent in 1954. In contrast to Lorenzo's tactics, Harris hoped to grow the airline, expand routes to places like Atlanta, and avoid cutting back on service or firing employees. While employees gladly welcomed this approach as "a breath of fresh air," Harris's tactics may have been ill-timed to attend to Continental's huge losses while the United States was engaged in the Persian Gulf War and also entering a recession. Losses kept mounting under Harris, who saw the company lose $314 million in the first half of 1991. Harris was forced out by August of that year and replaced with someone who respected financials: Robert R. Ferguson III, Continental's senior vice president for corporate development and a financial strategist with the company since 1985.

Ferguson's early days as CEO seemed promising, for he proposed a two-pronged strategy to save Continental by dealing with its past debts while simultaneously forging a new way to raise revenues in the immediate future. Ferguson's first task as CEO, however, was to shepherd the company through its second bankruptcy. Unlike the first Continental bankruptcy, there were no significant cash reserves to defray outstanding obligations. Ferguson solicited new investors to provide the necessary capital as he was splitting the company into two brands—CAL Lite focused on short haul, low frills service and Continental focused on traditional service for long-haul national and international flights. Ferguson proposed making half of Continental's 20,000 annual flights low fare CAL Lite offerings. This strategy seemed to Ferguson like the wave of the future in the new era created by deregulation. Reporters at *Business Week* were somewhat hopeful about the proposal: "CAL Lite will try to imitate the successful strategy of Southwest Airlines Co."[15]

Though his launch of CAL Lite may have proven to be an effective strategy in theory, Ferguson's overeagerness to get this airline within an airline up-to-speed quickly caused many problems. Ferguson, industry analysts believed, had expanded "Lite" too quickly. "As a result," commented the *New York Times*, "parts were not available for repairs at some airports, resulting in many canceled flights."[16] Although "Lite" service was supposed to brighten Continental's image as a poor service provider, its problems soon just confirmed consumers' perceptions of the airline as second-rate. After three years of losses with CAL Lite, Ferguson was pushed out of the leadership of Continental.[17]

When Ferguson stepped down as Continental CEO in October 1994, Bethune volunteered to be his replacement.[18] Continental's board agreed to give Bethune the job of CEO and to elevate Brenneman to Bethune's former position of COO. Bethune was not given much time between his hiring as CEO and his first performance evaluation. Within 10 days of becoming CEO, Bethune was required to present the board with an action plan for the company's short- and long-term horizons.[19]

Bethune's Path to Continental Airlines

Bethune was born in San Antonio, Texas, in 1941, to parents of modest means, who divorced while he was a youngster. Bethune lived with his mother in San Antonio for most of his childhood, and she tried to make ends meet by selling encyclopedias. Bethune credits his mother for being a fair but tough disciplinarian.[20] His father was the one who introduced Bethune to airplanes. Bethune's father flew crop dusting planes in Mississippi, and on a summer visit when Bethune was 15, Gordon helped his father land his crop duster each night at 8 p.m. on a makeshift landing strip. To prepare the plane's landing, Bethune put a smudge pot on the fence at the end of the landing strip and parked his father's 1950 Plymouth at the beginning, so the car's headlights would illuminate where his father could safely touch ground. The tasks appeared mundane, but Bethune understood that their importance to the overall success of his father's work day and career were quite significant:

> That summer I learned not only that when you have a responsibility, you'd better carry it out, but that there are a lot of jobs in which the consequences of screwing up were pretty drastic. Not just being a surgeon or the president, but being a pilot or mechanic—or the guy who put the smudge pot out to light the runway. Everybody else can do their job well—the mechanic gets the engine just right, the chemical guy loads up the plane, the pilot flies perfectly—but if the guy who's supposed to park the Plymouth at the end of the runway falls asleep or lets the battery die or steps out for a beer, well, it would be a pretty bad day for everybody.[21]

Bethune was less interested in his high school studies than going out to see the world, so, at age 17, Bethune decided to quit school and join the U.S. Navy. He remembers how he successfully managed that negotiation with his mother. "I went up to the attic and got the family suitcase and I started packing. My mom wanted to know what I was doing and I said, 'I'm going to California, and I'll send the suitcase back when I get there. I wanted to join the Navy but

I couldn't until I was 18 if you wouldn't sign the permission, so I'm going to hitchhike out to California and find some work, something to do until I'm 18 and then I'm going to join the Navy.' She said, 'Oh, no you're not, you're going to fall in with the wrong kind of people and possibly be in jail or get killed.' So I said, 'Well, Mom, if you really don't like that, I do have those papers that you can sign to let me join the Navy next week when I'm 17.' ... Five days after I turned 17, I joined the Navy."[22]

Bethune's first assignment in the navy was in California, working as an airplane mechanic. After passing the U.S. High School General Equivalency Test, Bethune was dissatisfied with having an equivalent degree, and after his transfer to Key West, Florida, Bethune enrolled in a local high school's evening program and earned a diploma. Early on in his naval career, Bethune showed a keen sense of how to stand out from the pack:

> The Navy is a huge, bureaucratic environment, and everybody gets a well-defined job. In that environment, guys just sort of do their jobs and disappear. I figured, how much better do I have to be in order to get the attention of my superiors and get the better assignments, the better jobs, the better promotions? And I figured only about 10% better. In other words, not much better. In an environment where people are happy to just do their jobs and stay out of trouble, making sure you do a little better, a little something extra, come to work a little more prepared, is enough to get you noticed and rewarded. So I started doing that.[23]

During Bethune's 20-year naval career, he earned licenses as an airframe and power plant mechanic and rose to the rank of lieutenant. Besides learning about how to build and repair planes, Bethune credits his superiors and teachers in the military for providing him training and opportunities to lead people. With some reluctance, Bethune agreed to manage his fellow navy mechanics when there was an opening for an evening supervisor at one of his stations. In his final year in the service, Bethune and his maintenance department were recognized as the best in the nation by the U.S. Navy's Chief Operations Office. "We won the award that year," Bethune would later say. "You know why? Get those guys to want to do it. And to want to do it you've got to respect them, and you've got to talk with them in their language."[24]

Upon leaving the navy just before airline deregulation, Bethune worked in executive positions at Braniff, Western, Piedmont, and Boeing before joining Continental. When Bethune started work at Braniff in 1978, it was the leading commercial airline in Dallas. In September of that year, Braniff announced that its third quarter earnings hit a record high of $15 million, an increase of 81 percent over the prior quarter.[25] In January 1979, Braniff offered its customers Concorde flights in Air France and British Airways planes out of the Dallas-Fort Worth Airport.[26] The good times for Braniff did not last. In 1979, Braniff's earnings declined by 21 percent in the second quarter, then, in the fourth quarter it suffered a $51.4 million loss.[27] As Braniff Chairman Harding Lawrence explained, the price spikes in international oil and fuel markets in 1979 were dampening the company's profitability.[28] Braniff was soon bankrupt, and Bethune cited the reasons for Braniff's demise: "The company had a bad business plan, high costs, poor management and, when all was said and done, a poor product."[29]

Bethune stayed at Braniff until 1983, learning more about the administrative side of the airline business while spending his evenings earning a bachelor's degree in business at Abilene Christian University in Dallas, before switching to Western Airlines and working out of its Los Angeles office. It was there that Bethune decided to get a private pilot's license, learning to fly a single engine Cessna. The following year, Bethune improved his flying skills after taking a position at Piedmont Airlines in North Carolina as senior vice president of operations by enrolling in Piedmont's training school: "That experience helps me when I'm dealing with pilots. We're part of the same club. I'm not just some suit. I'm a pilot, and I'm a mechanic. When we're talking about the best way to run an airline, they know that I know something. They respect what I've done. We speak the same language. If I say, 'Don't over rotate,' when I'm talking about not getting the company moving too fast, they get the idea."[30] After leaving Piedmont to work at Boeing, Boeing CEO Phil Condit urged Bethune to take up golf to improve his capacity to sell potential customers on Boeing products. Bethune disagreed and offered an alternative: he wanted to learn how to fly Boeing jets, believing that it would help his credibility in the eyes of Boeing employees. Bethune later reflected on the decision: "You have to prove that you're adding value to the team, that you know something, that you can do something, that you're going to be an asset."[31] Bethune became a certified pilot of Boeing 737s and 757s, the same planes manufactured by Boeing's Renton Division, the division Bethune managed.[32]

Bethune Takes Charge at Continental

In 1994 when Bethune stepped into the CEO role, it appeared that Continental would suffer the same fate as Pan Am. Like Pan Am, Continental had suffered years of losses along with frequent changes in uncertain leadership. Bethune did not believe that Continental's problems were terminal; things could be turned around, but the airline would require some radical surgery. Reflecting back on the state of Continental Airlines when he took over in 1994, Bethune described the company culture: it was a "deal culture" that had been created by Lorenzo 15 years before. Bethune was not a deal maker like Lorenzo, but he understood how they thought:

> Deal makers look at problems and think deals.... Did you ever see a deal maker try to fix a watch? It doesn't work. Deal makers buy and sell watches. If a watch is broken, they usually try to sell it. Or they buy more watches to divert attention from the broken watch—to bury its problems while they're making money on other watches. I did not think more deals were going to work. We're an airline and the solutions to our problems were airline solutions, not deals.[33]

At Continental, Bethune's first step was to create a plan that could transform the company from the inside out—from a deal culture to an airline culture. With the board's approval, Bethune and Brenneman unveiled the "Go Forward" plan in January 1995 to Continental's employees: Fly to Win (a product-service plan); Fund the Future (a financial plan); Make Reliability a Reality (the product plan); and Working Together (the people plan).[34] This bold analysis of Continental's present problems and future potential demonstrated to the airline's employees that this was a recovery plan with a chance of success. Gaining the good will of employees was the necessary first step in moving Continental in the right direction. Bethune's leadership style was critical to revising the sinking airline and getting employees to trust him was critical. He inspired them through his own experience as an airline employee and his ability to relate to them collectively was one of the most important aspects of the turnaround at Continental. Bethune was able to ask employees to make a renewed commitment after years of being ignored by management.

Fly to Win

Fly to Win was designed to allow Continental to maximize its strengths. According to Bethune, Brenneman presented the concept as a way of reestablishing a winning tradition at Continental. "We decided to stop doing things that didn't make money," Bethune said. "Greg said that at a meeting once and I really liked the sound of it. It sounds simple, even obvious. After all why would you do things that wouldn't make money?"[35] With this clear-headed approach, Bethune surveyed Continental's route system and determined that Continental's long-time hub, Denver, was losing money. Bethune decided to close Continental's operations in Denver and shift resources to more promising locales. As Brenneman put it, "we were going to build up our Houston, Newark and Cleveland hubs." In addition, we were going to increase revenues by expanding "our customer mix from backpacks and flip-flops to suits and briefcases."[36]

When Bethune joined the company in spring 1994, Continental planes flew 22 percent of the flights out of Denver, its most expensive hub to run. Meanwhile, Continental had 77.6 percent of all flights out of Houston, and nearly half of all flights out of Newark and Cleveland.[37] When Bethune explained his rationale to Continental employees, he did so by emphasizing that certain programs and people were cut not only to save money, but also to strengthen core businesses and the long-term health of the company. "We were losing money flying all of these point-to-point routes for ridiculously low fares in markets where we were at best a minor player. We decided to drop those routes and fly only to places it seemed people actually wanted to go. We realized that we simply had too many seats out there—like the farmer who tries to sell more tomatoes in a glutted market at depressed prices. Removing capacity would keep the same revenue while reducing our cost."[38]

Bethune and Brenneman crunched the numbers and determined that 18 percent of Continental's flights were not making money, and, that most of those were flights on CAL Lite.[39] Bethune and Brenneman saw CAL Lite as a drag on the Continental brand and jettisoned it from Continental's future offerings. Bethune noted that "frequent shuttle service in markets like New Orleans and Houston work well, and we'll continue to have that. It won't necessarily have 'Lite' painted on the side of the airplane and it won't be an all-coach airplane but it will be everything else. The part of 'Lite' that was profitable and works will continue. We don't necessarily have to call it a separate product."[40] The decision came at a great cost—a one-time $446.8 million charge added to first quarter 1995 losses of $534.2 million.

Continental's recent heritage of making its customers unhappy was another obstacle to recovery. Bethune met the problem head on by getting on the phone—along with many other Continental executives—and apologizing to once-loyal travel agents and business travelers for the airline's past sins. Furthermore, Bethune sought out customer feedback on Continental service. Keeping an eye on combining efficiency and good customer service, Bethune explained his approach: "We didn't ask, 'What do you want?' We asked, 'What would you be willing to pay for?' "[41] In response, Continental kept its OnePass frequent flier program, put first-class seats and food back in all Continental planes, and focused on flying customers to places they regularly wanted to fly.[42]

Fund the Future

On Thanksgiving 1994, almost a month after Bethune and Brenneman first presented the Go Forward plan to Continental's Board, Brenneman determined that Continental did not have sufficient cash reserves to pay its 40,000 employee salaries in January 1995. Bethune and Brenneman determined that Continental's planes comprised the most expensive fixed assets, so they came up with a three-pronged cost reduction plan based on restructuring Continental's fleet. First, Continental reduced the number of fleet types from 13 to 4. Second, Continental matched airplane sizes with the size of the consumer market. Third, Continental eliminated above-market leases on planes.[43] One of the first planes that Bethune phased out was the Airbus A300—Continental purged itself of all 21 of this class of plane by January 1995 and the 4,000 employees specifically designated to service them.[44] Bethune also approached Continental's most sympathetic creditors to relax the size and length of the repayment terms for existing loans.

These actions still did not generate enough savings nor were they timely enough to prevent Continental from another potential bankruptcy. In December 1994, Brenneman informed Bethune that the company would be forced to declare bankruptcy by the end of January without an immediate infusion of new funds. Bethune decided to try to find money wherever he could and begged Boeing, his former employer, to refund Continental's $70 million deposit for a new series of planes. It was not industry practice for a deposit to be returned, even if an order was canceled. However, Bethune decided to make a call for help to Boeing President Ron Woodard: " 'Ron,' I said, 'I know you're not contractually obligated to return that money for our cancelled orders but dammit, you need to give us this money back. We need it in the worst way.' Ron kind of laughed . . . then he thought about it for awhile. He trusted me. He believed in the direction we were pushing Continental. 'I'll tell you what,' he said. 'We'll give you half of it back.' He said he'd send it to me. I said, 'Ron, you don't understand. I need you to wire it to me.' Ron laughed. And he wired us $29 million."[45] According to Brenneman, that wire transfer allowed Continental to make payroll on January 17, 1995, and averted the bankruptcy crisis.[46] Here again, we see how the dogged personal commitment and resourcefulness of leaders can determine the fate of companies who survived in an industry.

Bethune later added that cost-cutting could not be the central component of funding the future. "Any dumb S.O.B. can manage costs. Revenue is a lot harder. You can make an airline so cheap nobody will fly it. It's not how cheap the product is; it's got to have some value," he stated.[47] That is one concept that Bethune believed separated his business philosophy from the ones practiced by his immediate predecessors: "My predecessors at Continental were more concerned about saving money than keeping customers. The old philosophy was that the only way to make money was to cut costs.... They actually rewarded pilots for saving fuel. Now I'm a pilot and I know how to save fuel. You just slow the airplane down and turn off the air conditioner. The pilots were up in the front thinking they were doing great, while all of our customers in the back were hot and late—not a good combination if you want to keep your customers happy."[48]

Make Reliability a Reality

Bethune and Brenneman decided to measure Continental's reliability using 15 performance metrics, the most significant and most public being those measured by the Department of Transportation. These included statistics for on-time performance, mishandled bags, overall customer complaints, and involuntary denied boardings.[49] To Bethune, airlines had to do a few things well: "Reaching destinations on time. Being clean, safe and reliable."[50] If there were obstacles to achieving those goals, Bethune needed to know so these obstacles could be addressed and eliminated. Realizing that there might be some reluctance on the part of employees to speak up out of fear of the repercussions, he set up an 800 telephone number that only he answered.[51]

To become more reliable, Continental also had to hear the latest feedback from another constituency—Continental's customers. Continental placed postage paid reply cards in in-flight magazines and asked fliers to grade services rendered. Bethune also created a customer 800 number, 800-WECARE2, to give customers a way to voice their concerns and get a response from corporate headquarters within five business days. Bethune discovered that letters of complaint sent to Continental in the past were often ignored, or if there was a response, it was a response along the lines of what Bethune humorously said: "Sorry you hated your flight. Hope you'll fly with us again!"[52]

Open communication inside the Continental corporation was another important aspect to Bethune's strategy to make the airline more reliable. Bethune kept his executive office open, ready to receive employees and their questions at any time during the day. He also held weekly open houses on Friday nights, inviting staff to meet and talk with him over drinks and snacks outside his office. This was considerably different from employees' interaction with senior management before Bethune took the helm. "You would have senior management in offices screaming for people that would be rows away," an employee recalled. "It was so disrespectful. People would be crying in their cubicles. It was horrible. It was nasty. You lost your self worth."[53]

Working Together

Bethune and Brenneman were convinced that mutual trust was required to make the Go Forward plan work, to get people working together rather than against each other in a sort of Darwinistic survival of the fittest culture. "Members of Continental's existing management team were not up to this challenge," Brenneman said. "They were too busy trying to knock each other off. In fact, for 15 years, the way to get ahead at Continental was to torpedo someone and take his or her job."[54] Two actions that Bethune initiated were the elimination of the intrusive security cameras at Continental's headquarters in Houston and the removal of locks to the executive suite.[55]

The Go Forward plan itself presented common company-wide goals for working together, but Bethune added a financial incentive for on-time flights to further encourage collaboration. If employees upped Continental's percentage of on-time flights over the previous quarter, then the company would distribute a special $65 bonus check monthly to each employee. When the Department of Transportation report came out in April 1995, Continental rose to an 80 percent rating, up 21 percent from the first quarter of 1994, and 35,000 employees each received a $65 pretax bonus for the next 3 months. When Continental improved its second quarter 1995 numbers, another set of checks for $65 were sent out to the entire company for 3 more months. Employee trust and confidence in Continental grew stronger with each day. Bethune believed that "there is a 100% correlation between employee happiness and customer satisfaction."[56]

Given that Continental was rising up the ranks of the major carriers, Bethune promised that in fiscal year 1996, if employees contributed to elevating the company to the top three Department of Transportation rankings for on-time flights, a $100 monthly bonus would be the reward. In February 1996, Continental was listed as the best major commercial airline in the United States for on-time flights in the fourth quarter of 1995. As a result, Bethune had a decision to make—the company had already exceeded the bonus goal for 1996, so, should employees get the promised $65 or the future bonus of $100? Bethune opted to give everyone at Continental the $100 reward for finishing first in the final quarter of 1995.[57] Empowering employees was part of what Bethune hoped would create innovation and incremental improvement. As Bethune said, "you have to take up all of those good ideas you and your cronies come up with at meetings and actually put them into practice."[58]

In early 1995, Bethune made another break from the Lorenzo-Ferguson era by inviting his employees into the company parking lot in Houston to join him in a ceremonial burning of the old Continental employee manuals and rule books. At the center of the bonfire was what was known around Continental as "The Thou Shalt Not Book," a bible almost nine inches thick stuffed with dictums about what Continental employees were allowed to do and what was forbidden.[59] Staff was asked to throw their copies on top of the bonfire.

> Symptomatic of an organization in which nobody trusted anybody, we had rules—specific rules—for everything from what color pencil had to be used on boarding passes to what kind of meals delayed passengers should be given to what kind of fold ought to be put in a sick-day form. Even worse, it spelled out job responsibilities to such a fine degree that employees were utterly bound by arcane rules and demands, and the penalties for disobeying the rules were severe. If a person whose plane was cancelled had an unrestricted full-fare ticket, that person might get a hotel room; an adjoining passenger with a less expensive ticket might get a meal voucher and a pat on the arm. It didn't matter if this policy started a war at the airport. The gate agent was not allowed to solve a problem that didn't make sense, so he or she just had to take the abuse the passengers gave and smile. The smiling, needless to say, had pretty much stopped.[60]

In a sense, the Continental rule book was a concrete manifestation of a dominant business model that worked when the industry was growing under the comfortable protection of regulation. Standards in such a setting helped to eliminate variance and provided clear guidelines, but when the firm needed to be more nimble and responsive, these standards became part of an iron cage of bureaucracy.

Bethune worked with his senior management team to identify the strong players, the role players, the staff members who need improvement, and those who were an overall drag on the achievement of team goals. Bethune ranked the strong players highest with a I, the weakest, the lowest ranking of IV. Those ranked as a IV would be encouraged to take an early retirement or find employment elsewhere. Bethune argued that his decision to let go employees he ranked as a IV was not a setback. Remaining Continental staff realized that these people could not be part of the new way of doing business and would be better off elsewhere. "All through 1995 we talked to people about their performance, giving them a chance to either get on board with the company's new direction or not, as they saw fit. I don't say it was easy, because you never want to fire employees... [But] that final cut didn't cause the smallest amount of unhappiness or fear or dissatisfaction in the ranks. Instead, there was a big sigh of relief. Employees said, 'Jeez, they got rid of Harry—that jerk should have been shot 20 years ago and somebody finally did something.' "[61]

According to Brenneman, team leadership was the driving philosophy behind the Go Forward plan. "Cultivating honesty, trust, dignity and respect becomes the job of the leaders. It may even be their most important job; Gordon and I certainly considered it our top priority."[62] "Bethune's formula is deceptively simple," Jonathan Burton noted in *Chief Executive* magazine. "He restored first-class seats and personally courted top customers. Thinly traveled markets were cut. He replaced 34 of 60 vice presidents with two dozen handpicked corporate doctors, focused on revenue, cost, and margins. For the first nine months of 1995, Continental earned $183 million. And a $6,500 investment in its Class B shares in January 1995 was worth $36,000 just 10 months later."[63]

Gordon Bethune at Houston Intercontinental Airport in 1996. (Source: Getty Images).

Continental's Upward Trajectory

As Bethune and Continental implemented the Go Forward plan, accolades soon followed. In June 1996, J. D. Power named Continental as the "Best Airline with Flights of 500 Miles or More," in essence, the best airline in the major commercial airline industry. *Air Transport World* was equally impressed, and honored Continental as its airline of the year for 1996. Later in 1996, Bethune was identified by *Business Week* as one of the top 25 leaders in the world that year. In 1997, the Smithsonian Institution and the trade journal *Aviation Week & Space Technology* honored Bethune with its Laureate in Aviation trophy.

To enhance Continental's service to its customers, Continental announced on November 20, 1998 that it had formed an alliance with Northwest Airlines that translated into a reciprocal relationship on code-sharing on domestic and international routes, reciprocity of frequent flyer programs and shared airport lounges. The alliance expanded Continental's industry status to be on par with the big three U.S.-based airlines—Delta, United, and American. When Continental issued its annual report for the year 2000, it highlighted five of the strongest years in company history. Between 1995 and 2000, Continental's revenue nearly doubled, from $5.8 billion in 1995 to $9.9 billion in 2000.[64] Bethune's renewed focus on serving his business travelers had a sizeable impact on revenues—by 2000, 47.6 percent of all Continental's revenues were from business travelers, up from 32.2 percent in 1994.[65]

Bethune and Continental after September 11, 2001

Continental seemed poised for even more growth in the twenty-first century, but like all other airlines, it was ill-prepared for the after-effects of the terrorist attacks on the United States on September 11, 2001. In the immediate aftermath of the attacks, President George W. Bush ordered all commercial planes grounded for the next day and a half. Although this caused a significant loss of revenue, Continental decided to give full refunds to all passengers with tickets to fly on those days, and then pledged to fly them again at no cost once the president restored normal airline services. Continental lost millions of dollars each day—as much as $30 million a day starting on September 12—and few people were willing to fly, compounding the revenue lost with the refund policy.[66] Layoffs looked like a necessary evil to stem losses. Rather than put 1,000 workers on furlough, Bethune offered a range of options to his employees, including a new Company Offered Leave of Absence option. "The COLA encourages voluntary leaves of absence for non-management employees and allows them to accept full-time work elsewhere while continuing to receive benefits during an unpaid leave of up to one year. Benefits include the option to continue insurance benefits at company rates, continuing travel privileges and credit for time on COLA for company seniority and the employee retirement program."[67]

This program translated into a $60 million loss to Continental, but Bethune expressed the negative impact in terms of the loss of valued human resources. "This furlough was one of the most painful events we have had to experience," Bethune said. "We believe that employees should always be treated with dignity and respect, especially when we are forced to make these tough decisions."[68] To amplify this point, Bethune and COO Larry Kellner (Brenneman left in 2000 to become CEO of Burger King), decided to forgo all of their personal compensation for the rest of 2001.

On September 21, 2001, President Bush addressed Congress and the nation, offering words of solace and reassurance about America's future and declared that the government would expand its role in airline and transportation safety and create a new Department of Homeland Security: "We will come together to improve air safety, to dramatically expand the number of air marshals on domestic flights, and take new measures to prevent hijacking. We will come together to promote stability and keep our airlines flying, with direct assistance during this emergency."[69] To make that possible, Bush signed an executive order pledging $15 billion to the airline industry, $5 billion in direct federal

Former U.S. President George Bush (C) walks through Bush Intercontinental Airport with Continental Airlines CEO Gordon Bethune (L) after coming home to Houston aboard a Continental jet from Boston September 27, 2001. Bush and Bethune held a brief press conference at the airport to emphasize the importance of consumer confidence in airline safety. (Source: Reuters/CORBIS).

aid and $10 billion in loan guarantees for an industry that had announced tens of thousands of layoffs since the terrorist hijackings.[70]

Continental managed to avoid a major loss in 2001, and in 2002, continued to build on each of the four parts of its Go Forward plan. In January 2002, Continental partnered with Amtrak to provide air/train ride share service in the Northeast United States. In March 2002, Continental opened its new jet terminal in Newark and broadcast the availability of its BusinessFirst ticket program to frequent fliers on Boeing 777 planes. Later that year, Continental's Houston hub opened its Airport of the Future with a computerized and automated check-in program, soon to be implemented at airports throughout the nation. As Continental was taking these steps to rebuild customer confidence and sales in the wake of 9/11, another disaster struck—Continental's fourth major hub in Guam was devastated by super typhoon Ponsogna, contributing to a $109 million fourth quarter loss and a net loss of $451 million for 2002, Continental's first significant annual loss since 1994.[71]

Unlike 1994, the losses were not connected to customer dissatisfaction and low worker morale. In 2002, for the fifth year in a row, *Conde Nast* magazine rated Continental best among all U.S. airlines in comfort, reliability, and value on transatlantic and transpacific flights. For the sixth year in a row, Continental was listed in *Fortune* magazine's top 100 Places to Work.

Despite doing more with less, Bethune decided with Kellner that Continental had to take further austerity measures in Spring 2003, trimming senior management positions by one-quarter, the top officer group by 15 percent and issuing 1,200 furloughs to save $500 million. Overall, the airlines had lost $300 billion since the September 11th attacks, and, with the country's offensive in Iraq, it was expected that travel would drop and fuel prices would increase, a double strike against Continental's already lackluster financial figures.

In September 2003, Bethune announced EliteAccess, as a new way to cut into the discount flier market. Believing that "there are still people who want creature comforts," Bethune offered full-fare customers perks like no middle seats, priority boarding and baggage handling, and faster security checks. These services, already available to Continental's business-class and first-class customers, were extended to economy-class customers who paid full fare, typically one-fifth of all economy-class customers.

Bethune agreed to step down at the end of 2004, and from May 2003 to May 2004, Bethune and Continental squeezed out $900 million in operating income improvements to offset the additional $700 million Continental would expend on fuel due to the oil price surges. He resisted any attempt to renege on Continental's financial obligations to current staff and retirees, putting $103 million in cash and another $100 million into the pension fund in September 2003. It was a markedly different approach than that taken by US Airways (the successor to US Air), which terminated its pilot pension plan, and United, which was lobbying Congress for a bailout or a relaxation of regulatory requirements for pension programs at U.S. corporations.[72] With great reluctance, in December 2004, Bethune's last month as CEO, Continental announced a minor pay cut, the last U.S. airline to take this step. However, Kellner, Bethune's successor as CEO, and Jeff Smisek, Kellner's replacement as second in command, did the same, accepting more than 20 percent less in compensation.

As Kellner and Smisek assumed their new roles at Continental, they took stock of Bethune's Go Forward plan and decided that it would remain the guiding vision of their company. "I don't think you'll see a transition in regard to strategy, in regard to culture," Kellner said. "This will continue to be all about our employees and our customers. And if we treat our employees well and do the right thing there, they will treat our customers well, [and] we aren't going to lose our focus on our people and our product."[73]

A New Model for the Major Carriers

Bethune understood, better than Trippe or Lorenzo, what constituted success in the modern era of airline business—an era in which competitors fiercely cut costs and customers demanded high-quality service. The pioneering days of Pan Am, when flying a rickety Ford Tri-Motor was an unparalleled thrill for many adventurous passengers, were long gone. Gone also were the heady days of airlines under regulation that contributed to safe, accessible, and reliable air travel for a broader array of consumers. The pursuit of profits in the early years of deregulation, when Lorenzo's financial wizardry created short-term shareholder gains at the expense of sound operations management, was an unsustainable model for running Continental or any of the other major carriers. Consumers now had many choices concerning which airlines they would fly.

While Herb Kelleher and Southwest Airlines pursued a tightly focused and integrated business model based on inexpensive, no frills point-to-point service in the era of deregulation, Bethune and Continental Airlines attempted to pursue a customer-oriented business model that could work within the strictures of the hub and spoke model favored by the major historical carriers in the airline industry. Bethune's Continental opted for more creature comforts and better service to attract the most lucrative customer segment—business travelers. In many ways, Bethune's approach to turning around the airline was centered on a "back to basics" plan, which involved a streamlining of unprofitable sectors and a corresponding investment in services that customers wanted and were willing to pay for. The beauty of Southwest's and Continental's business models is the seamless alignment that connects incentives and benefits for employees directly to customer service. Looking at the state of the airline industry in the twenty-first century, it would appear that Southwest has created one new model of success. The "back to basics" approach that Bethune and Brenneman championed to revitalize Continental may very well become a winning formula for the traditional, large-scale carriers.

Epilogue

Throughout this book, we have used the airline industry to demonstrate the manner in which leaders can influence and are influenced by the context of their times. This reciprocal relationship between the individual and his or her environment can result in the creation of new businesses (by entrepreneurs), the expansion of opportunities (by managers), or the rebirth of companies on the decline (by leaders). Beyond creating, expanding, or reenergizing businesses, entrepreneurs, managers, and leaders can have a significant influence on an industry's evolution. Although there has been much written about industry evolution, the literature has often neglected the role that leaders can and do play in shaping that evolution. Through the stories of founders and CEOs in the airline industry during the twentieth century, we hoped to highlight the powerful impact that leaders have had on an industry's lifecycle. We believe that the manner in which airline CEOs and founders have shaped their industry is indicative of the way that executives of other industries influence their competitive and contextual landscape.[1]

The co-evolution of context and individual action not only significantly shapes the structure of an industry, it can also provide the foundation for a dominant business model as was the case in the second phase of the airline industry's evolution. As the lifecycle of an industry evolves, this dominant business model must also evolve if it is to retain its relevance.[2] The actions of individuals can reinforce a dominant business model by building barriers to entry through solidifying favorable legislative protection, shaping consumer demand, or influencing technology deployment. Alternatively, individuals can seek to challenge the dominant business model and influence industry evolution by carving out niche opportunities or specific target consumer segments as Pacific Southwest Airlines (PSA) did. In some cases, entrepreneurial activity within a period of stability and maturity may create the foundation for the "next" dominant niche business model. PSA's model was adopted and adapted by Southwest Airlines which, in turn, became a new and profitable business model—one that established airlines have tried to emulate. Likewise, opportunities that emerge from reinvention and turnarounds championed by leaders, such as Gordon Bethune's transformation of Continental, can also redefine the future parameters for success and alter the competitive landscape. The actions of leaders not only influence the dominant business model in an industry, but they can also lengthen or shorten the periods of dominance in an industry's lifecycle. For instance, Juan Trippe's tireless lobbying efforts to secure and defend a global monopoly on international air travel enabled Pan American World Airways to sustain a dominant position in the industry for almost four decades.

As we have seen in our stories of airline executives, entrepreneurs played key roles in building a viable industry during a period of great uncertainty and confusion. During start-up periods, there are typically several models of potential success and a variety of competitive frameworks—each one vying for dominance. Beyond start-up phases, entrepreneurs have also played a key role throughout the industry by introducing new potential models of success. Some start as small niche opportunities that then became the foundation for a dominant business model design while others carve out a perfectly comfortable position as an alternative, albeit a nondominant business model.

As an industry grows to critical mass, managers take a prominent role in defining a dominant business model that enables companies to dramatically expand their scale and scope. Managers shepherd the industry through massive periods of growth by scaling their operations and introducing innovations to protect and sustain their dominance. The level of variation in the competitive landscape during periods of maturity typically narrows. Managers focus on how to operate better, faster, and cheaper within the dominant business design. In some respect, the actions of managers during periods of industry maturity can have an unexpectedly significant impact on the industry's evolution. The innovations that they introduce to differentiate their businesses during this time are typically adopted by all other mainstream competitors. Differentiation is thus a fleeting advantage, as its rapid imitation simply recalibrates the base level of success for the entire industry. During the regulated, mature phase of the airline industry's evolution, C. R. Smith of American Airlines and Pat Patterson of United Air Lines tried to differentiate their companies through their introduction of customer loyalty programs, stewardesses, and on-site food preparation—however, all quickly became standard practices in the industry.

The cresting of an industry's demand growth provides opportunities for leaders or change agents to develop new models that can fundamentally redefine the conditions and parameters of success. Through a focus on streamlining, operational efficiency, or bold reinvention, creative leaders can establish a new or reestablish a dominant business model as Gordon Bethune did with Continental Airlines. These times typically call for a significant focus on reengineering existing business models and processes. This can also be an opportunity to exploit pockets of latent potential as Herb Kelleher and Southwest Airlines did when they targeted nonflying regional commuters.

Certain periods during an industry's lifecycle seem to be particularly poised for specific leadership archetypes (entrepreneurs during start-ups, managers during periods of growth and maturity, and leaders during periods of decline). Yet, across all phases, an industry develops through a co-evolutionary process in which individuals shape and are, in turn, shaped by the context of their times. The balance of power between the individual and the environment can vary over time (see table E.1). During certain periods, individuals have a

Table E.1 How airline executives shaped and were shaped by their context

	How executive was shaped by context	*Executive's role in shaping context*
Harry Guggenheim *Fund for the Promotion of Aeronautics*	• Role of airplanes in World War I • European airline industry dominance • High profile accidents • Technology advances by Wright and Curtiss	• Public perception • Safety • Sharing of technical information • Promotion of Charles Lindbergh
C. E. Woolman *Delta Air Lines*	• Air show in France in 1909 • Locked out of early airmail contracts • Market opportunity in agriculture • Underserved regional market • Service to country during World War II	• Combined agribusiness with aviation • Created regional service • Started passenger service before airmail service • Move to hub and spoke model after World War II
Juan Trippe *Pan American World Airways*	• Glut of World War I surplus planes • Locked out of domestic airmail contracts • World War II—need for expansion in Pacific • Postwar international competition (IATA) • Expansion of international competition post-World War II	• Competed against slower international travel alternatives • Heavy government lobbying • Technological advances (jets) • Military partnership • International diplomacy – "ambassador" for U.S. interests around globe
C. R. Smith *American Airlines*	• Part of "Big 4" industry consolidation—beneficiary of Postmaster Brown • Regulation • Fear of flying	• Introduction of DC-3 • Safety focus • SABRE reservation system • Customer loyalty programs • Admirals Club • Flight training center
Pat Patterson *United Air Lines*	• Part of "Big 4" industry consolidation—beneficiary of Postmaster Brown • Regulation • New aircraft	• Stewardesses • Airline food service • Centralized maintenance and operational centers • Labor-management relations
Herb Kelleher *Southwest Airlines*	• Success of Pacific Southwest • Less stringent intrastate legislation • Deregulation	• Alignment of systems, technology, culture, and people • Single aircraft—standardized maintenance and support • Low cost, no frills • Attract new customers (non-air travelers) • Point-to-point service
Gordon Bethune *Continental Airlines*	• Deregulation • Intense competition • Leveraged buyouts • Mergers & acquisitions • Industry consolidation • Demoralized staff/industry	• Focus on business customers • Back to basics approach • Streamlining, simplifying operations • Elimination of bureaucracy

disproportionate impact on the evolution of the industry. This was certainly the case during the start-up phase of the airline industry. During other periods, the context defines the parameters for success so tightly that it constrains the latitude of an individual's actions.

Lessons Learned

Industries are shaped by the actions of both internal and external forces. Individuals as the agents of internal change can establish, protect, and redefine a dominant business model of success at a particular point in time. External forces such as government regulation, technological innovation, or geopolitical forces can define the contextual landscape and have an equally large effect on the evolution of business models in an industry. Of course, such external factors can also be influenced by the actions of individuals (through lobbying, investments in technological partnerships, and global alliances). By understanding the nature of this co-evolutionary process, we are in a better position to determine the type of individuals and the leadership approaches that are best aligned for success. Our analysis of the co-evolutionary process in the airline industry in the twentieth century has yielded a number of important lessons for any industry.

1. *Aligning leadership approaches to the appropriate industry lifecycle*—As industries move from one stage of development to the next, there is often a need to reevaluate the leadership approaches that have been historically successful. During the shake-out phase of the early airline industry, there was plenty of room for a wide variety of business approaches and entrepreneurial activity. In fact, many entrepreneurs competed with each other to establish a business model that would be profitable and sustainable. When the dominant business model emerged, the nature and role of leadership needed to change. There was less focus on new development and experimentation and more focus on protection and optimization. Certain individuals can make this leap from entrepreneur to manager, but it can be extremely difficult for many who feel stifled by the closing down of opportunities and possibilities and the increasing need for focusing on standards and building a stable organization. It is important to be attuned to the lifecycle stage of an industry to determine the type of leadership that is required for success.
2. *Impact that individuals have on influencing the direction of the industry*—As is evidenced in table E.1, individuals can influence the direction of an industry to a significant degree during any stage of an industry's lifecycle. For instance, in the early years of the airline industry, Harry Guggenheim helped to promote the growth of the industry through investments in technology, information sharing, and public relations, and Juan Trippe played a significant role in defining the conditions and options for international travel originating from the United States. C. R. Smith was a key player in establishing passenger travel as a

viable economic engine through his work with Douglas Aircraft to create the DC-3, the first commercially viable passenger plane. Trippe played a similarly influential role in the advent of the jet age in the early 1960s. During the postregulatory phase of the airline industry's evolution, Gordon Bethune of Continental Airlines adopted a back to basics approach to streamline the company's operations for a radically changed competitive landscape. Though industries are shaped by several environmental factors, including demographic shifts, social mores, and government intervention or nonintervention, individuals have opportunities to harness these forces for their own advantage and the advantage of their company. As we have seen, many of these external forces can also be heavily affected by the actions of the individuals themselves.

3. *Co-evolutionary process of leadership style and context and the need for adaptation*—The co-evolutionary development of an industry essentially requires individuals to adapt and change their approaches as the context and conditions for success change. Many managers entrenched in the protection of a dominant business model lose sight of the changing competitive landscape and fail to modify their approach. This failure can not only lead to their dismissal, but can also lead to the ultimate demise of the company. Industry dominance can be a double-edged sword. Protecting one's position within an industry is only half the battle. It is important for managers who are in a dominant industry position to be cognizant of the competitive forces or environmental factors that can reshape the competitive landscape. A hyperfocus on one way of doing business can prevent even longer-term success. Too many contextual factors change for a business model to remain largely immune forever. If an industry remains the same for too long, the underpinnings for that stability can become artificial or forced. If a company chooses to ignore the percolating changes on the contextual horizon, they risk losing their relevance. Many mature airlines who invested in one model of success were fundamentally unable to move quickly enough to build a new format for success when the contextual landscape changed. This failure to adapt was, in large part, the death knell for Pan Am and Eastern.

The lessons learned from the CEOs and founders in this book are not unique to the airline industry; they can be applied to the study of any industry. For instance, the dramatic implosion of the financial services sector in 2008 is an indication that a new type of leader and a new form of leadership is required. The industry lifecycle in financial services has crested and is on the downward slope. The laissez-faire government policies in this sector and the growth at all costs approach of business leaders are no longer relevant. Complex alliances, obscure financial packaging, and excessive compensation have been replaced with a need for greater openness and transparency. The future focus of leaders in this industry will be less about growth and more about change, cost containment, and operational efficiency—all within the new strictures of government oversight. Beyond orchestrating the turnaround, leaders will need to reconceptualize and reinvent the financial services sector. The automobile industry is facing a similar fate and will require leaders to shepherd its turnaround.

While the financial services and automobile industries are facing turmoil and upheaval, many web-based businesses are facing the initial challenges of growth and maturity having passed through the entrepreneurial phase. Amazon, Google, Yahoo, and e-Bay have already been engaged in characteristically managerial activities, including mergers, acquisitions, alliances, and vertical integration. As these businesses seek to expand and capitalize on their initial growth, they are focusing more on the scale and scope of their operations. E-Bay has focused on acquiring various entities in the online delivery and payment chain. The growth of others such as Google will depend on the expansion of services (i.e., increase in scope) beyond its core search engine. And Yahoo's survival may very well depend on a strategic alliance or merger with a larger entity. Individuals at the helm of these web-based businesses will need to have a broader managerial outlook to ensure that their dominance within the industry is protected and ultimately, sustained.

As these web-based businesses work on the optimization and maximization of their business models, entrepreneurs are scrambling to define a business paradigm in the green/environmental sector. Although there have been several previous attempts to launch a sustainable green industry, no company has been able to gain significant traction to develop a dominant business model. That appears to be changing in the face of global warming and the constant calls for renewal energy. In many ways, former Vice President Al Gore is the Guggenheim of the green revolution. Like Guggenheim, Gore has played the role of evangelist in trying to galvanize public support and political backing for green-based industries. The commercial and consumer climate seem well poised to support entrepreneurs who can deliver "green" technologies, products, and services. Over the next several years, a variety of businesses will likely emerge to capitalize on these opportunities.

These stories demonstrate that it is important for any individual to understand the contextual framework in which he or she operates and how that framework relates to the lifecycle stage of the industry. Are there pockets of new opportunity? Is there a dominant, entrenched business model for success? If so, how does one penetrate it? In what ways can the context be shaped to create a new opportunity or to reinforce a dominant position in the industry? The answers to these questions will help individuals to assess what type of leadership is required and what opportunities are on the horizon. The answers will also demonstrate the degree to which individuals can influence their own outcomes. We hope that our stories of airline executives inspire readers to be both respectful and aware of the power of environmental factors and yet feel empowered by their personal ability to influence their context.

Notes

Introduction

1. "Conventional Wisdom," *Newsweek*, June 17, 2002.
2. James C. Collins, *Good to Great: Why Some Companies Make the Leap and—Others Don't* (New York: HarperBusiness, 2001).
3. For a discussion on the link between CEO pay and firm performance, see Rakesh Khurana, *Searching for a Corporate Savior: The Irrational Quest for Charismatic CEOs* (Princeton, NJ: Princeton University Press, 2002), pp. 20–50.
4. Over the century, there have been multiple studies of industry evolution based on a variety of perspectives, including sociological, population ecology, dominant design theory, lifecycle theory, economic evolution, stakeholder theory, and organizational alignment. The Bibliography section of this book includes a listing of some of the major historical studies of industry evolution. These studies are cited throughout the book.
5. Anthony J. Mayo and Nitin Nohria, *In Their Time: The Greatest Business Leaders of the 20th Century* (Boston: HBS Press, 2005), p. xv.
6. Joseph A. Schumpeter, *Capitalism, Socialism and Democracy* (Cambridge, MA: Harvard University Press, 1961), p. 74.
7. For more on Schumpeter's life and research see Thomas K. McCraw, *Prophet of Innovation: Joseph Schumpeter and Creative Destruction* (Cambridge, MA: Belknap Press of Harvard University Press, 2007).
8. Alfred D. Chandler, Jr., *The Visible Hand* (Cambridge, MA: Harvard University Press, 1977), p. 7.
9. Warren Bennis and Burt Nanus, *Leaders: The Strategies for Taking Charge* (New York: Harper & Row, 1985), pp. 17–18.
10. In their study of organizational evolution, Michael Tushman, William Newman, and Elaine Romanelli note: "In the emergence phase of a product class, competition is based on product innovation and performance, where in the maturity stage, competition centers on cost, volume, and efficiency. Shifts in patterns of demand alter key factors for success." In essence, the primary drivers of successful leadership need to change as an industry evolves. Michael L. Tushman, William H. Newman, and Elaine Romanelli, "Convergence and Upheaval: Managing the Unsteady Pace of Organizational Evolution," *California Management Review* 29 no. 1, 1986, p. 36.
11. Within *In Their Time*, we specifically looked at the context of distinct decades in the twentieth century and though they are not static, the decades provide natural

markers that displayed unique characteristics. We noted: "Though certain periods of the last century seem almost aligned for the emergence and dominance of a particular type of executive (e.g. the entrepreneur in the early twentieth century, the 'organization man'/manager in the post-WWII fifties, and the change agent in the tumultuous seventies), we found that all three types co-exist and are pervasive through every decade. Beyond being present in each decade, we found that all three archetypes were necessary and vital to sustaining the vibrancy of the capitalist system. Entrepreneurs create new businesses, managers grow and optimize them, and leaders transform them at critical inflection points. The cycle of the American capitalist system has borne witness to this business lifecycle over and over, and it is the ongoing regeneration of this cycle that ultimately sustains development and progress." Mayo and Nohria, *In Their Time*, pp. xxv–xxvi.

12. Mathew L. A. Hayward, Dean A. Shepherd, and Dale Griffin, "A Hubris Theory of Entrepreneurship," *Management Science* 52 no. 2, February 2006, p. 160. The authors note: "When founders are overconfident, they overestimate the likelihood that their ventures will succeed and that they can ensure such success." For more on hubris and overconfidence see: Nathan J. Hiller and Donald C. Hambrick, "Conceptualizing Executive Hubris: The Role of (Hyper-) Core Self-Evaluations in Decision-Making," *Strategic Management Journal* 26 (2005). Hiller and Hambrick note on page 307: "Not only is self-confidence thought to be substantially valuable, because it allows the manager to create and seize opportunities, as well as obstacles; but the external appearance of self-confidence is also valuable, for its motivational power and its potential to reassure constituents that the organization is in capable hands." Even Joseph Schumpeter, who famously heralded the entrepreneur's role as the "rogue elephant" that remakes the economy through a process of "creative destruction," emphasized that entrepreneurship is "a feat not of intellect, but of will. It is a special case of the social phenomenon of leadership. Its difficulty consisting in the resistances and uncertainties incident to doing what has not been done before, it is accessible for, and appeals to, only a distinct type which is rare." Joseph A. Schumpeter, "The Instability of Capitalism," *Economic Journal* 38 no. 151, September 1928, pp. 379–380. See also Joseph A. Schumpeter, *The Theory of Economic Development: An Inquiry into Profits, Capital, Credit, Interest, and the Business Cycle* (Cambridge, MA: Harvard University Press, 1934).
13. Hayward, Shepherd, and Griffin, "A Hubris Theory of Entrepreneurship," p. 163.
14. Mark Morrison, "Herb Kelleher on the Record, Part 3," *Business Week* Online, December 24, 2003, available at www.businessweek.com.
15. In the early phase of an industry, institutional pressures of the type identified by Meyer and Rowan (1977) and DiMaggio and Powell (1983) that drive conformity across organizations in an industry are still weak. Indeed, the personalities of individuals are almost a more powerful influence on the organization than the economic or institutional environment. In contrast to Stinchcombe (1960) who argues that organizations that are founded at a certain moment bear a common imprint that reflects the prevailing economic and institutional circumstances, we find that the imprint of the environment in this early phase is superseded by the imprint of the entrepreneur's idiosyncratic vision for their company. As a result many organizational models coexist, each vying for dominance. See John W. Meyer and Brian Rowan, "Institutionalized Organizations: Formal Structures as Myth and Ceremony," *American Journal of Sociology* 83 no. 2, September 1977, pp. 340–363; Paul J. DiMaggio and Walter W. Powell, "The Iron Case Revisited: Institutional Isomorphism and Collective Rationality in Organizational Fields," *American*

Sociological Review 48 no. 2, April 1983, pp. 147–160; and Arthur L. Stinchcombe, "The Sociology of Organization and the Theory of the Firm," *Pacific Sociological Review* 3 no. 2, Autumn 1960, pp. 75–82.

16. As industries mature, entrepreneurs search for niche opportunities or other ways to supplant a dominant business model as Herb Kelleher did to a large extent with Southwest Airlines. In some cases, these actions result in a new dominant business model. Clayton Christensen's and Joseph Bower's work on the combination of rapid technological change and organizational inertia suggest why many large firms can be particularly vulnerable to entrepreneurial energy and unpredictability. Most large firms, the authors contend, require impetus from large customers to innovate. Unfortunately, for established firms, "[b]ecause the rate of technical progress can exceed the performance demanded in a market, technologies which initially can only be used in emerging markets later can invade mainstream ones, carrying entrant firms to victory over established companies." Over time, the niche business models outpace the dominant players and in some cases, define a new level of dominance. Clayton M. Christensen and Joseph L. Bower, "Customer Power, Strategic Investment, and the Failure of Leading Firms," *Strategic Management Journal* 17 no. 3, 1996, p. 197.
17. The role of the managerial archetype is best described by the historian Alfred D. Chandler, Jr. See Alfred D. Chandler, Jr., *Strategy and Structure: Chapters in the History of the Industrial Enterprise* (Cambridge, MA: MIT Press, 1962) and Chandler, Jr. with Takashi Hikino, *Scale and Scope: The Dynamics of Industrial Capitalism* (Cambridge, MA: Belknap Press, 1990).
18. Michael L. Tushman reinforces this notion by stating that "the key role for executive leadership during convergent periods [periods of industry stability] is to reemphasize strategy, mission, and core values and to keep a vigilant eye on external opportunities and/or threats." Tushman, Newman, and Romanelli, "Convergence and Upheaval," p. 40. Tushman states that unlike periods of convergence, "frame-breaking change, however, requires direct executive involvement in all aspects of the change."
19. Tushman, Newman, and Romanelli, "Convergence and Upheaval," p. 35. The authors note: "Organizational history is a source of tradition, precedent, and pride which are, in turn, anchors to the past. A proud history often restricts vigilant problem solving and may be a source of resistance to change." When faced with environmental threat, organizations with strong momentum: may not register the threat due to organization complacency and/or stunted external vigilance (e.g. the automobile or steel industries), or if the threat is recognized, the response is frequently heightened conformity to the status quo and/or increased commitment to "what we do best." See also L. Richard Harrison and Glenn R. Carroll, *Culture and Demography in Organizations* (Princeton, NJ: Princeton University Press, 2006), p. 231. Harrison and Carroll state: "One of the well-known debates within organization theory concerns the adaptability of organizations. Some theories assume or imply that formal organizations are highly adaptive, while others contend that successful adaptation is rare and problematic, preferring instead an image of organizations as highly inertial."
20. The idea of the transformational leader, espoused by scholars such as John Kotter, Warren Bennis, and Howard Gardner which almost became the defining image of the leader in the late twentieth century, was very much inspired by executives who took charge of once dominant companies in decline and revived them to some semblance of their former glory. Perhaps the archetype of such leaders was Lee Iacocca

in business and Ronald Reagan and Margaret Thatcher in government. See John P. Kotter, *A Force for Change: How Leadership Differs from Management* (New York: Free Press, 1990); Bennis and Nanus, *Leaders: The Strategies for Taking Charge*; and Howard Gardner with Emma Laskin, *Leading Minds: An Anatomy of Leadership* (New York: BasicBooks, 1995).

21. Michael T. Hannan and John Freeman note that "managers or dominant coalitions scan the relevant environment for opportunities and threats, formulate strategic responses, and adjust organizational structure appropriately." Michael T. Hannan and John Freeman, "The Population Ecology of Organizations," *American Journal of Sociology* 82 no. 5, March 1977, p. 930.
22. Case studies can serve many purposes (Yin). Besides the general benefit of inductive theorizing, some cases have the benefit of being critical natural experiments in helping illuminate a theory. A critical natural experiment is ideally one in which the conditions make it unlikely for the main theoretical argument to hold. See Richard K. Yin, "The Case Study Crisis: Some Answers," *Administrative Science Quarterly* 26 no. 1, March 1981, pp. 58–65. For example, if one finds labor mobility to be hazardous even in a context in which workers are prima facie expected to be fully mobile (see Boris Groysberg's work on the mobility of sell-side investment analysts), it is a critical case for a theory that emphasizes the importance of organizational affiliation for the performance of any kind of worker. See Boris Groysberg, Andrew N. McLean, and Nitin Nohria, "Are Leaders Portable?" *Harvard Business Review*, May 2006. Similarly, we believe the airline industry is a critical case because the environmental forces have been so strong, that it should favor theories like population ecology (Hannan and Freeman) that suggest that individual CEOs can have but a limited influence on the evolution of an industry, in contrast to our theory which suggests that individuals can be as important as the environment in shaping the evolution of an industry. See Hannan and Freeman, "The Population Ecology of Organizations."
23. Anita McGahan, "How Industries Change," *Harvard Business Review*, October 2004. See also Paul R. Lawrence and Davis Dyer, *Renewing American Industry* (New York: Free Press, 1983), p. 22.
24. David K. Onkst, "Early Exhibition Aviators," *U.S. Centennial of Flight Commission: History of Flight*, http://www.centennialofflight.gov/essay/Explorers_Record_Setters_and_Daredevils/early_exhibition/EX7.htm (accessed June 30, 2006).
25. McGahan and Kou note: "Throughout the 1920s, entry was easy—all it took was an airplane, a pilot, a ticket office, and a competitive mail bid." Anita McGahan and Julia Kou, "The U.S. Airline Industry in 1995," HBS Case No. 795–113 (Boston: Harvard Business School Publishing, 1995), p. 1.
26. "Men Like Lindbergh Needed," *Washington Post*, October 25, 1927.
27. Robert J. Serling, *Eagle: The Story of American Airlines* (New York: St. Martin's Press, 1985), p. 35.
28. Willis Emmons notes: "As a result of entry and price controls, competition in the industry centered on service quality, including frequency of flights, aircraft type, and in-flight amenities." Willis Emmons, *The Evolving Bargain: Strategic Implications of Deregulation and Privatization* (Boston: Harvard Business School Press, 2000), p. 30. Ghemawat and Donohue note: "With suppressed route and price competition, the airlines focused their marketing ploys on service items such as meals and movies and challenged one another rigorously in this area." Pankaj Ghemawat and Nancy Donohue, "The U.S. Airline Industry, 1978–1988 (A)," HBS Case No. 390–025 (Boston: Harvard Business School Publishing, 1989), p. 2.

29. George Thomas Kurian, *Datapedia of the United States 1790–2005* (Lanham, MD: Bernan Press, 2001), pp. 335–336.
30. Roger E. Bilstein, "C.R. Smith: An American Original," in *Airline Executives and Federal Regulation*, ed. W. David Lewis (Columbus, OH: Ohio State University Press, 2000), p. 96.
31. Anita McGahan, "How Industries Change," *Harvard Business Review*, October 2004. McGahan notes that industries are not changed "overnight"; it often "takes decades for change to become clear and play out."
32. Richard H. K. Vietor, *Contrived Competition: Regulation and Deregulation in America* (Cambridge, MA: Harvard University Press, 1994), p. 70.
33. Ibid., p. 82.
34. Ibid., p. 13.
35. For more on People Express see Michael Beer and Philip Holland, "People Express Airlines: Rise and Decline," HBS Case No. 490–012 (Boston: Harvard Business School Publishing, 1993).
36. JayEtta Z. Hecker, "Commercial Aviation: Structural Costs Continue to Challenge Legacy of Airlines' Financial Performance," Government Accountability Office, July 13, 2005 (GAO-05–834T), p. 3.
37. Kurian, *Datapedia of the United States*, pp. 335–336. The spike in the number of U.S. airline operators between 1945 and 1950 reflects the anticipated increase in passenger travel following World War II. The increase in passenger traffic did not occur as swiftly as anticipated, and as a result, this expansion caused a short-term decrease in the profitability of many major airlines. As the market stabilized, a process of rationalization and consolidation occurred that leveled the competitive landscape until the mid-1970s.
38. Yann Cochennec, "The Profitability Problem," *Interavia* 55 (June 2000), pp. 28–29.
39. Hecker, "Commercial Aviation," p. 6.
40. External forces like the shock of the terrorist attacks on the United States on September 11, 2001 can result in a fundamentally different industry structure. Greiner calls these industry jolts, "*periods of revolution* because they typically exhibit a serious upheaval of management practices." Larry E. Greiner, "Evolution and Revolution as Organizations Grow," *Harvard Business Review*, May–June 1998. Donald N. Sull calls these jolts "sudden-death threats" noting that "these are major environmental shocks that can put a company out of business in relatively short order." Donald N. Sull, "Strategy as Active Waiting," *Harvard Business Review*, September 2005. Much of the previous research on industry changing jolts has been focused on the role of innovation and technology. See Daniel A. Levinthal, "The Slow Pace of Rapid Technological Change: Gradualism and Punctuation in Technological Change," *Industrial and Corporate Change* 7 no. 2, 1998, p. 217. Levinthal explicitly uses the theory of punctuated equilibrium to describe the relationship between incremental/gradual changes and more massive or "discontinuous" changes in the field of technology.
41. Hecker, "Commercial Aviation," p. 7.
42. Air Transport Association of America, "Annual Revenue and Earnings: U.S.Airlines" available at http://www.airlines.org/economics/finance/Annual+US+Financial+Results.htm (accessed July 11, 2007).
43. Clayton M. Christensen, Scott D. Anthony, and Erik A. Roth, *Seeing What's Next: Using the Theories of Innovation to Predict Industry Change* (Boston: Harvard Business School Press, 2004). See chapter 6 for a discussion of the potential future of the airline industry.

Part I The Entrepreneurs

1. Scott Bruce and Bill Crawford, *Cerealizing America* (Boston: Faber & Faber, 1995) and Gerald Carson, *Cornflake Crusade* (New York: Rheinhart, 1957).

Chapter 1 The Guggenheims: Promoting Aviation in America

1. "Guggenheims to End Fund for Aviation," *New York Times*, October 29, 1929.
2. "Aviation is Weaned," *The Commonweal* 11 (November 13, 1929), p. 34.
3. Ibid.
4. Harry F. Guggenheim, *The Seven Skies* (New York: G. P. Putnam's, 1930), p. 96.
5. See Donald S. Lopez, *Aviation: A Smithsonian Guide* (New York: Macmillan, 1995). Although a few good pieces had been written on Guggenheim preceding the publication of the guide, they seem not to have persuaded the Smithsonian to include Guggenheim next to colleagues such as Lindbergh and Doolittle. For an excellent short article on the Guggenheim Fund see, for example, Richard P. Hallion, "Daniel and Harry Guggenheim and the Philanthropy of Flight," in *Aviation's Golden Age: Portraits from the 1920s and 1930s* (Iowa City: University of Iowa Press, 1989), pp. 18–34.
6. Dominick A. Pisano, "New Directions in the History of Aviation," in *The Airplane in American Culture*, ed. Dominick A. Pisano (Ann Arbor: University of Michigan Press, 2003), p. 7.
7. Guggenheim, *The Seven Skies*, p. 209.
8. Guggenheim news release, "Status of Civil Aviation," January 17, 1926, Daniel Guggenheim Fund Papers, box 1, Library of Congress, Washington, D.C., quoted in Richard P. Hallion, *Legacy of Flight: The Guggenheim Contribution to American Aviation* (Seattle, WA: University of Washington Press, 1977), p. 35.
9. Richard P. Hallion, *Taking Flight: Inventing the Aerial Age from Antiquity through the First World War* (Oxford: Oxford University Press, 2003), pp. 290–292.
10. Historian Richard P. Hallion describes how Curtiss attracted customers in the early days of aviation: "In 1917, by which time he already had a main plant, three branch factories, and training schools in New York, Virginia, and California, he issued a slickly produced and illustrated catalog of his products offering 'new worlds to conquer' for 'those whose enthusiasm for outdoor sports has, successively, led them through motoring, motor boating, hydroplaning, and ballooning.' " See Hallion, *Taking Flight*, p. 288.
11. Ford quoted in Hallion, *Taking Flight*, p. 291; discussion of the Wright and Curtiss feud in Hallion, *Taking Flight*, pp. 290–294 and 480–481 fn. 53.
12. Hallion, *Taking Flight*, p. 292.
13. Ibid., pp. 53–54.
14. Ibid., pp. 250–257
15. Ibid., pp. 383–384.
16. Ibid., pp. 355, 390. For more information on the S.E. 5a, see "One of the World's Finest Fighters," *Arkansas Air Museum*, http://www.akairmuseum.org/se5a.html (accessed September 13, 2005).
17. Hallion, *Taking Flight*, p. 388.
18. Ibid., pp. 389–390.
19. Irwin Unger and Debi Unger, *The Guggenheims: A Family History* (New York: HarperCollins, 2005), p. 247.
20. "Hoover," in *Aviation's Golden Age*, pp. 126–127.

21. This quotation was included in Harry Guggenheim's collection of writings on aviation, *The Seven Skies*, p. 102.
22. Hallion, *Legacy of Flight*, p. 23.
23. Ibid., pp. 24–25. See also "Harry Frank Guggenheim," *Dictionary of American Biography, Supplement 9: 1971–1975*. Charles Scribner's Sons. 1992, reproduced in *Biography Resource Center* (Farmington Hills, MI: Thomson Gale, 2005), http://galenet.galegroup.com.ezp1.harvard.edu/servlet/BioRC.
24. Daniel Guggenheim to Harry Guggenheim, 1918, quoted in Hallion, *Legacy of Flight*, p. 26.
25. Daniel Guggenheim quoted in Hallion, *Legacy of Flight*, p. 29.
26. For the military's skepticism about the potential of air power, see John F. Shiner, "Benjamin Foulois," in *Aviation's Golden Age*, p. 77. See also Hallion, *Legacy of Flight*, p. 4.
27. See "William P. MacCracken," in *Aviation's Golden Age*, p. 41.
28. Hallion, *Legacy of Flight*, p. 14.
29. These recommendations were also adopted in the Navy Five Year Aircraft Program and the Army Five Year Aircraft Program that were both passed by Congress in June 1926. For a discussion, see Hallion, *Legacy of Flight*, pp. 13–15.
30. Hallion, *Legacy of Flight*, p. 31.
31. Daniel Guggenheim quoted in Milton Lomask, *Seed Money: The Guggenheim Story* (New York: Farrar, Straus, 1964), p. 83.
32. Morrow was appointed by Coolidge as ambassador to Mexico in 1927. Morrow's daughter soon afterward married Charles Lindbergh.
33. Lomask, *Seed Money*, pp. 84–85.
34. "Daniel and Harry Guggenheim," in *Aviation's Golden Age*, pp. 22–23.
35. Daniel Guggenheim to Hoover, January 16, 1926, "Aviation—Daniel Guggenheim Fund," Commerce Papers, Hoover Library, quoted in "Daniel and Harry Guggenheim," *Aviation's Golden Age*, p. 153, fn. 12.
36. Unger and Unger, *The Guggenheims*, p. 250.
37. For a description of the trip, see Hallion, *Legacy of Flight*, pp. 37–38.
38. "Harry Guggenheim Will Push Air Plan," *New York Times*, April 29, 1926.
39. Hallion, *Legacy of Flight*, pp. 17–18.
40. Although Fokker had a manufacturing plant in the United States, he also licensed the F-7 to plants in seven other countries. See Pamela Feltus, "Fokker and His Aircraft," http://www.centennialofflight.gov/essay/Air_Power/Fokker/AP7.htm. Ford Motor Company came out with one of the first trimotor airplanes, the 2-AT Pullman, in early 1925. Their first completely successful version of the trimotor was the 4-AT, nick-named the "tin goose," which debuted on June 11, 1926. Ford's lagging behind the accomplishments of the Dutch-born Fokker was only one more indicator of European dominance in aviation. See "Ford Trimotor," http://www.centennialofflight.gov/essay/Dictionary/Ford_Trimotor/DI84.htm.
41. Hallion, *Legacy of Flight*, pp. 39, 128–130.
42. "$150,000 in Prizes for Safer Planes," *New York Times*, June 19, 1926.
43. Ibid.
44. Ibid.
45. Hallion, *Legacy of Flight*, pp. 101–102.
46. One of the most notable fatalities was that of French ace Charles Nungesser, who was lost over the Atlantic without a trace just two weeks before Lindbergh's triumph. See "Charles Nungesser: French Ace of WWI, 43 Victories," *Acepilots.com*, http://www.acepilots.com/wwi/fr_nungesser.html (accessed September 18, 2005).

47. "Guggenheim Cables Praise," *New York Times*, May 12, 1926.
48. Byrd began the first leg of the trip between Washington and New York but declined to complete the tour because of various speaking engagements.
49. "Byrd Plane Ends Tour of Country," *New York Times*, November 24, 1926.
50. For a detailed description of the contest rules, see Hallion, *Legacy of Flight*, pp. 130–132.
51. Hallion, *Legacy of Flight*, p. 130.
52. Guggenheim wrote: "It is evident that one disastrous crash in which two national heroes lose their lives has more effect on the public mind than a million miles of safe commercial flying which is more or less nationally invisible." Harry F. Guggenheim, "Safety in the Air," *The Saturday Evening Post*, June 25, 1927, p. 29.
53. Ibid., p. 160.
54. Ibid., p. 29.
55. Ibid., p. 169.
56. For a detailed account of Lindbergh's meteoric rise on the world stage, see A. Scott Berg, *Lindbergh* (New York: Putnam's, 1998), pp. 135–177.
57. Many of the early articles by Lindbergh could be found on the front page of the *Times*. See, for instance, "Lindbergh Says Flying Boats Will Come in 5 to 10 Years," *New York Times*, June 9, 1927. For an example of his series of articles, see his third, which focused on safety—perhaps influenced by Guggenheim's example: "Lindbergh on Flying: The Third Article Discusses the Safety Quest," *New York Times*, September 9, 1928.
58. Harry married his second wife Caroline Morton Potter in 1923. In the spring of that year, the couple traveled to France with an architect to gather ideas about designs for a house and to buy furniture. Falaise was based on a Norman farmhouse and was completed in 1924. It was used principally as a summer home. Unger and Unger, *The Guggenheims*, pp. 240–241, 244–245.
59. "Tour by Lindbergh Brought to an End: Seen by 30,000,000," *Washington Post*, October 23, 1927.
60. Harry F. Guggenheim, "Commercial Flying Can Be Made Safe," *Forbes*, July 25, 1927, p. 9.
61. Ibid., pp. 9–11, 38.
62. "Guggenheim Demands Curb on Stunt Flights," *Washington Post*, September 8, 1927. Guggenheim's lack of finesse in trying to change forcibly the mores of his fellow Americans is reflected in the unusually opaque and inarticulate prose Guggenheim employs to make his case.
63. One measure of the clout Harry Guggenheim enjoyed within the airline industry was that the group of executives who came to New York included some of the biggest names in aviation, such as the Canadian financier Clement Keys from National Air Transport and William B. Mayo from the Ford Motor Company. See Hallion, *Legacy of Flight*, p. 88.
64. Hallion, *Legacy of Flight*, pp. 87–88.
65. For information on the collaboration between Hanshue and Guggenheim, see Patricia A. Michaelis, "Harris M. Hanshue," in *Encyclopedia of American Business History and Biography: The Airline Industry*, ed. William M. Leary (New York: Facts on File, 1992), pp. 205–206; Robert J. Serling, *The Only Way to Fly: The Story of Western Airlines; America's Senior Air Carrier* (Garden City, NY: Doubleday, 1976), pp. 55–76; and Hallion, *Legacy of Flight*, pp. 86–95.
66. "Guggenheim Aids Flying Progress," *New York Times*, September 2, 1928.
67. Hallion, *Legacy of Flight*, p. 111.

68. Lindbergh quoted in Harry Guggenheim, "Aviation—Progress in Safety," *Harvard Business Review,* October 1929, p. 40.
69. Unger and Unger, *The Guggenheims*, p. 261.
70. Harry Guggenheim, "Giving Wings to the World," *St. Nicholas* (December 1928); and idem, "Making Flying Safe," *Forum* (July 1929).
71. At this time, Guggenheim was on the verge of announcing a winner. In January 1930, the Fund awarded $150,000 to Curtiss Tanager. Although the Tanager airplane was soon after destroyed in a hanger fire, the legacy of the competition survived in the development of techniques for "short-takeoff-and-landing" (STOL). According to one historian, the competition helped the Fund to "formulate the requirements for... [STOL] aircraft long before many in the aeronautical community realized such a need existed. Without question, the modern STOL airplane today, with its profusion of high-lift devices, owes its conception in part to the Guggenheim competition of 1927–29." See Hallion, *Legacy of Flight*, pp. 149–150.
72. Guggenheim, "Aviation—Progress in Safety," p. 41.
73. Ibid., pp. 37, 40–41, 43.
74. "Finds Americas Field for Air Expansion," *New York Times*, September 8, 1929.
75. T. J. C. Martyn, "When Aviation Arrives at Mass Production," review Guggenheim, *The Seven Skies* in *New York Times*, September 14, 1930.
76. Unger and Unger, *The Guggenheims*, p. 265.
77. Martyn, "When Aviation Arrives at Mass Production."

Chapter 2 Juan Trippe's Early Entrepreneurial Efforts

1. Marylin Bender and Selig Altschul, *The Chosen Instrument: Pan Am, Juan Trippe—The Rise and Fall of an American Entrepreneur* (New York: Simon and Schuster, 1982), p. 87.
2. Trippe's appreciation of the international implications of mechanical flight may help to explain why Charles Lindbergh and he worked so closely together for so long during the early years of Pan American Airways. Trippe certainly would have agreed with this assessment by Lindbergh in 1930: "Aviation must be considered from an international standpoint. An ability to cover great distances in a relatively short time makes it a leading factor in world intercourse." Charles Lindbergh quoted in David L. Butler, "Technogeopolitics and the Struggle for Control of World Air Routes, 1910–1928," *Political Geography* 20 (2001), p. 635.
3. Portions of this chapter are drawn from: Anthony Mayo and Nitin Nohria, *In Their Time: The Greatest Business Leaders of the 20th Century* (Boston: Harvard Business School Press, 2005), pp. 92–97, and Nitin Nohria, Anthony J. Mayo, and Mark Rennella, "Juan Trippe and Pan American World Airways," HBS Case No. 406–086 (Boston: Harvard Business School Publishing, 2006).
4. Bender and Altschul, *The Chosen Instrument*, p. 31. See also, "Wright Plane Flies Over the Bay," *New York Times*, September 30, 1909.
5. Nohria, Mayo and Rennella, "Juan Trippe and Pan American World Airways," p. 1.
6. Bender and Altschul, *The Chosen Instrument*, pp. 49–50.
7. Robert Daley, *An American Saga: Juan Trippe and his Pan Am Empire* (New York: Random House, 1980), p. 8.
8. Bender and Altschul, *The Chosen Instrument*, p. 59.
9. An historian of the aircraft industry writes: "Production [in the United States] shrank from 14,020 aircraft in 1918, to 780 in 1919 and to a post-war low of 263 in 1922." See G. R. Simonson, "The Demand for Aircraft and the Aircraft Industry, 1907–1958," *Journal of Economic History* 20 (1960), p. 365.

10. R. E. G. Davies, *A History of World's Airlines* (London: Oxford University Press, 1964), pp. 5–6. Henry Smith reveals the cause of the airline failure: "when tourists went north in the spring, the Tampa air line disappeared." Henry L. Smith, *Airways: The History of Commercial Aviation in the United States* (New York: Alfred A. Knopf, 1942), p. 84. See also W. David Lewis and Wesley Phillips Newton, *Delta: The History of an Airline* (Athens, GA: University of Georgia Press, 1979), p. 8. Despite its geographical advantages, the St. Petersburg-Tampa airline was still burdened by the primitive state of airplane technology, which kept the reliability of airplanes low.
11. Daley, *An American Saga*, p. 9.
12. Bender and Altschul, *The Chosen Instrument*, pp. 60–61 and Daley, *An American Saga*, pp. 9–11.
13. Daley, *An American Saga*, pp. 10–13.
14. Nohria, Mayo and Rennella, "Juan Trippe and Pan American World Airways," p. 2.
15. Daley, *An American Saga*, p. 13.
16. Davies, *A History of the World's Airlines*, p. 39.
17. Mayo and Nohria, *In Their Time*, p. 91.
18. Davies, *A History of the World's Airlines*, p. 42.
19. Robert J. Serling, *Eagle: The Story of American Airlines* (New York: St. Martin's Press, 1985), p. 8.
20. Theodore O. Wallin, "The Air Commerce Act Creates a Federal Airways System," in *Great Events from History II: Business and Commerce*, ed. Frank N. Magill (Pasadena, CA: Salem Press, 1994), p. 499.
21. Kenneth M. Morris, Marc Robinson, and Richard Kroll, *American Dreams: One Hundred Years of Business Ideas and Innovation from the Wall Street Journal* (New York: Lighthouse Press, 1992), p. 74. The section on the Kelly Act was originally included in: Anthony J. Mayo and Laura G. Singleton, "C. R. Smith and the Birth of American Airlines," HBS Case No. 406–082 (Boston: Harvard Business School Publishing, 2006), pp. 2–3.
22. Daley, *An American Saga*, pp. 14–15.
23. That Trippe encountered the awesome power of federal government regulation over the airline industry in 1925 probably contributed to his ability to take advantage of the U.S. government's interest in promoting its international agenda through the intermediary of a private American airline in the late 1920s.
24. Daley, *An American Saga*, pp. 14–15 and Bender and Altschul, *The Chosen Instrument*, p. 66.
25. Daley, *An American Saga*, pp. 14–16.
26. Ibid., p. 20.
27. Wesley Phillips Newton, "Juan T. Trippe," in *Encyclopedia of American Business History and Biography: The Airline Industry*, ed. William M. Leary (New York: Facts on File, 1992), p. 467. Aeromarine Airways got its start in 1919 by flying passengers from Key West to Cuba who sought to briefly escape the restrictions of prohibition in the United States. The airline also operated charter flights from New York City to the Great Lakes region during the summer months, but was not able to secure a long-term postal subsidy. Over the course of 3 years, the airline flew more than 30,000 passengers, but a fatal crash in 1923 forced the struggling airline out of business. See William M. Leary, "Aeromarine Airways," in *Encyclopedia of American Business History and Biography: The Airline Industry*, ed. William M. Leary (New York: Facts on File, 1992), pp. 7–8.
28. Newton, "Juan T. Trippe," p. 467. Trippe is reported to have told one of Colonial's major investors Robert Thach, a lawyer who met Trippe during his time in the

U.S. Navy, to "fix it [the composition of the board] so that our crowd is in control." See also Bender and Altschul, *The Chosen Instrument*, p. 70.

29. Pamela Feltus, "Fokker and His Aircraft," in *U.S. Centennial of Flight Commission: History of Flight*, http://www.centennialofflight.gov/essay/Air_Power/Fokker/AP7.htm (accessed July 13, 2005).
30. "Fokker History—Long Version," *Fokker Services*, http://www.fokkerservices.com/page.html?id=5402 (accessed July 15, 2005).
31. Feltus, "Fokker and His Aircraft."
32. Judy Rumerman, "Wright Aeronautical Company," in *U.S. Centennial of Flight Commission: History of Flight*, http://www.centennialofflight.gov/essay/Aerospace/Wright_Aero/Aero8.htm (accessed July 15, 2005).
33. Bender and Altschul, *The Chosen Instrument*, p. 70.
34. Daley, *An American Saga*, pp. 17–19 and Bender and Altschul, *The Chosen Instrument*, p. 70.
35. Daley, *An American Saga*, p. 16.
36. Ibid., pp. 16, 472.
37. Ibid., p. 20 and Newton, "Juan T. Trippe," p. 467.
38. Trippe wrote this assessment of O'Ryan to a banker with interests in Colonial: "Inasmuch as I have no confidence whatsoever in his ability to successfully manage an air transport company I am resigning as vice president." Colonial did eventually succeed at providing service for CAM 1 and was soon bought by a company that would later be acquired by American Airlines. See Bender and Altschul, *The Chosen Instrument*, pp. 76–77.
39. The name for Trippe's corporation can easily be confused with Aviation Corporation, incorporated in 1929 and discussed in the following chapter. Depending on the sources being used, both corporations have been referred to by the acronym AVCO. However, works on Pan Am usually do not use that acronym for Juan Trippe's Aviation Corporation of America, while works on American Airlines do call its parent company AVCO.
40. Daley, *An American Saga*, p. 9.
41. For information on Harriman, see Rudy Abramson *Spanning the Century: The Life of W. Averell Harriman, 1891–1986* (New York: Morrow, 1992). For Beckers, see Bender and Altschul, *The Chosen Instrument*, p. 96.
42. Bender and Altschul, *The Chosen Instrument*, pp. 78–79.
43. Newton, "Juan T. Trippe," pp. 467–468.
44. Bender and Altschul, *The Chosen Instrument*, p. 82.
45. Ibid., p. 82.
46. See R. E. G. Davies, *Delta, an Airline and Its Aircraft: The Illustrated History of a Major U.S. Airline* (Miami: Paladwr Press, 1990), pp. 13, 17.
47. See the article and accompanying map in "Flyers' Own Story of Hop to Moscow," *New York Times*, June 26, 1931.
48. Whitney quoted in Bender and Altschul, *The Chosen Instrument*, p. 82.
49. Trippe was not a warm person, but he was very adept at turning on the charm when it suited his interests. Historian Wesley Phillips Newton succinctly describes this shrewd aspect of Trippe's character: "Trippe was dark, stocky, taciturn, aggressive to the point of arrogance, but careful around elders who could advance him. One of Trippe's most persuasive characteristics was his unexpected smile, which could both captivate and intimidate." See Newton, "Juan T. Trippe," p. 464.
50. Trippe's biographers are unsure exactly where, when, or how he managed this meeting. Robert Daley speculates that they met at a dinner held for Lindbergh by

the mayor of New York a day after the ticker-tape parade welcoming the young pilot. See Daley, *An American Saga*, p. 478. However the meeting was arranged, Trippe no doubt used his extensive social network in New York along with his own experiences as a pilot to meet and then connect with Lindbergh. That the two young men also shared similar visions about the airplane's potential to have a huge impact on international commerce and relations must have helped Trippe to overcome Lindbergh's awkwardness during this frenzied time in his life.

51. Juan Trippe, interview by Robert Daley, quoted in Daley, *An American Saga*, p. 478.
52. Daley, *An American Saga*, p. 62.
53. Bender and Altschul, *The Chosen Instrument*, p. 83.
54. On the history of dirigibles, see Judy Rumerman, "The Era of the Dirigible," in *U.S. Centennial of Flight Commission: History of Flight*, http://www.centennialofflight.gov/essay/Lighter_than_air/dirigibles/LTA9.htm. On the history of the Zeppelin, see "The Zeppelin" in *U.S. Centennial of Flight Commission: History of Flight*, http://www.centennialofflight.gov/essay/Lighter_than_air/zeppelin/LTA8.htm. For the range of Germany's dirigibles, see Butler, "Technogeopolitics," p. 641. For a recent article that offers a brief overview of improvements in steamship travel and some of its cultural implications for American travelers see Mark Rennella and Whitney Walton, "Planned Serendipity," *Journal of Social History* 38 no. 2 (winter 2004).
55. Butler, "Technogeopolitics," p. 641.
56. Ibid., pp. 639–644. It was not until 15 years later that the United States made a similar declaration to that of Britain's in 1911. That declaration was made in the Air Commerce Act of 1926, which "affirmed the sovereignty of the United States over its airspace." See Bender and Altschul, *The Chosen Instrument*, p. 124.
57. Benedict Crowell quoted in Butler, "Technogeopolitics," p. 650.
58. Butler, "Technogeopolitics," pp. 648–649.
59. Ibid., p. 649.
60. Ibid., p. 646. Instead of welcoming this broad international treaty, the United States entered into discrete bilateral agreements concerning air travel. See "International Civil Aviation," in *U.S. Centennial of Flight Commission: History of Flight*, http://www.centennialofflight.gov/essay/Government_Role/Intl_Civil/POL19.htm (accessed July 27, 2005).
61. H. S. Bacon, letter of May 14, 1920 to Benedict Crowell, U.S. National Archives, quoted in Butler, "Technogeopolitics," p. 650.
62. Wesley Phillips Newton, *The Perilous Sky: U.S. Aviation Diplomacy and Latin America, 1919–1931* (Coral Gables, FL: University of Miami Press, 1978), pp. 25–27, 62–64. For a succinct history of SCADTA see Davies, *A History of the World's Airlines*, pp. 71–75.
63. Bender and Altschul, *The Chosen Instrument*, p. 84; Newton, *The Perilous Sky*, pp. 74–75, 109–114. Besides helping to spur the creation of Pan Am, Peter Paul von Bauer's legacy to American aviation may have been the section of the 1926 Air Commerce Act prohibiting foreign planes to fly through U.S. airspace without permission from the State Department. See Bender and Altschul, *The Chosen Instrument*, p. 141.
64. Newton writes that Postmaster "New publicly announced, following a cabinet meeting in May 1925, that the government would not support any project for a proposed airline connecting the United States and Panama unless it were undertaken by a company incorporated in the United States" with U.S. capital, personnel and planes. Newton, *The Perilous Sky*, p. 65.

65. Newton reports that Coolidge had been swayed in 1925 to oppose SCADTA's bid to service the Panama Canal publicly "after European military attachés in Washington sounded a warning that it would be unwise to let a German company establish air bases in Central America close to the Panama Canal." Newton, *The Perilous Sky*, p. 65.
66. James H. Smith, Jr. to Marilyn Bender, June 6, 1980, quoted in Bender and Altschul, *The Chosen Instrument*, p. 154.
67. Daley, *An American Saga*, p. 31. For a brief history of the Curtiss-Wright Corporation, see Judy Rumerman, "The Curtis-Wright Corporation," in *U.S. Centennial of Flight Commission: History of Flight*, http://www.centennialofflight.gov/essay/Aerospace/Curtiss_wright/Aero9.htm (accessed August 1, 2005).
68. Newton, "Juan T. Trippe," p. 468.
69. Bender and Altschul, *The Chosen Instrument*, p. 85.
70. Daley, *An American Saga*, p. 33.
71. Bender and Altschul, *The Chosen Instrument*, p. 86.
72. Nohria, Mayo and Rennella, "Juan Trippe and Pan American World Airways," p. 6.
73. Bender and Altschul, *The Chosen Instrument*, p. 87.
74. Ibid., pp. 92–93.
75. Historian Wesley Phillips Newton makes these inferences about how Pan American's aggressive plans for growth had an influence on the interdepartmental committee of 1927: "the provision giving the postmaster general leeway to elect the most desirable contractor was designed to prevent a contract from going to any company with the kind of ties that Peter Paul von Bauer offered; moreover, the confident attitude and actions of PAA before the passage of the act strongly [imply] that this provision had a second motive: the selection of a single 'chosen instrument' to carry out the government's aim of protecting the Caribbean through domination of the air routes in the vicinity of that artery." Newton, *The Perilous Sky*, pp. 161–162.
76. Bender and Altschul, *The Chosen Instrument*, pp. 92–93.
77. Allan Spalding, "Baron of the Airways," *New Republic* 110 (June 5, 1944), p. 768, quoted in Newton, *The Perilous Sky*, p. 159.
78. Bender and Altschul, *The Chosen Instrument*, p. 73.
79. Glover quoted in Bender and Altschul, *The Chosen Instrument*, p. 94.
80. Trippe quoted in Bender and Altschul, *The Chosen Instrument*, p. 95.
81. Bender and Altschul, *The Chosen Instrument*, pp. 95–96. By 1930 about a half dozen German and French-owned (or affiliated) airlines—such as the French Aeropostale and the German Kondor-Varig—had already been established all over South America. See Davies, *A History of the World's Airlines*, pp. 70–77.
82. Bender and Altschul, *The Chosen Instrument*, pp. 99–100.
83. Ibid., pp. 127, 134, and 140.
84. See Newton, *The Perilous Sky*, p. 161 and Bender and Altschul, *The Chosen Instrument*, p. 96. Although Trippe did not like all the provisions of the Act of 1928, he was adept at nudging government officials to bend the rules that, in his opinion, unfairly diminished Pan Am's profits. See, for example, Bender and Altschul, *The Chosen Instrument*, pp. 98–99.
85. Richard V. Oulahan, "Observations from Times Watch-Towers: Assess Hoover's Trip," *New York Times*, January 6, 1929.
86. This trip may have also been taken to portray Pan Am in a good light during a rivalry with an American competitor for South American air routes, NYRBA.
87. "Lindbergh Finishes Tropical Mail Flight," *New York Times*, September 24, 1929.
88. "Times World Wide Photos," *New York Times*, September 29, 1929.

89. Butler, "Technogeopolitics," p. 652.
90. Lindbergh quoted in Bender and Altschul, *The Chosen Instrument*, p. 98.
91. Bender and Altschul, *The Chosen Instrument*, p. 98.
92. Ibid., p. 113.
93. Ibid., p. 81. For a concise history of Hoyt's accomplishments, see "Richard F. Hoyt, Financier, 46, is Dead," *New York Times*, March 8, 1935.
94. Bender and Altschul, *The Chosen Instrument*, p. 105.
95. For the instance of preparing the way for Pan Am's servicing of the west coast of South America, see Bender and Altschul, *The Chosen Instrument*, p. 117.
96. Bender and Altschul, *The Chosen Instrument*, pp. 112–113 and Daley, *An American Saga*, p. 66.
97. Bender and Altschul, *The Chosen Instrument*, pp. 118–120.
98. Rowe quoted in Bender and Altschul, *The Chosen Instrument*, p. 100.
99. For Glover's antagonistic stance toward NYRBA, see Bender and Altschul, *The Chosen Instrument*, pp. 170–171.
100. The information on Pan Am's feud with NYRBA was summarized from an extended discussion in Bender and Altschul, *The Chosen Instrument*, pp. 166–176.
101. For Pan Am's first scheduled passenger service see, Asif Siddiqi, "Pan American: The History of America's 'Chosen Instrument' for Overseas Air Transport," in *U.S. Centennial of Flight Commission: History of Flight*, http://www.centennialofflight.gov/essay/Commercial_Aviation/Pan_Am/Tran12.htm. Morgan quoted in Bender and Altschul, *The Chosen Instrument*, p. 125.

Chapter 3 C. E. Woolman and Delta Air Lines

1. W. David Lewis and Wesley Phillips Newton, *Delta: The History of an Airline* (Athens, GA: University of Georgia Press, 1979), p. 5.
2. "15,000 See Airships Put Through Paces," *New York Times*, August 21, 1909.
3. Geoff Jones, *Delta Air Lines: 75 Years of Airline Excellence* (Chicago: Arcadia Publishing, 2003), p. 9 and Lewis and Newton, *Delta*, p. 7.
4. "Democratic Party Platform of 1920," http://www.presidency.ucsb.edu/showplatforms.php?platindex=D1920.
5. "C.E. Woolman Is Dead at 76; Chairman of Delta Airlines," *New York Times*, September 12, 1966.
6. Lewis and Newton, *Delta*, p. 7.
7. For a detailed description of Woolman's personal characteristics, see Lewis and Newton, *Delta*, pp. 30–36.
8. Henry L. Smith, *Airways: The History of Commercial Aviation in the United States* (New York: Alfred A. Knopf, 1942), pp. 47–48.
9. Lewis and Newton, *Delta*, pp. 11–12. By the end of 1920, according to Roger Mola, "there were 145 municipally owned airports and the nationwide airport system was beginning to form." See Roger Mola, "The Earliest Airports," in *U.S. Centennial of Flight Commission: History of Flight*, http://www.centennialofflight.gov/essay/Government_Role/earliest_airports/POL9.htm (accessed April 13, 2005). The improvements to the aircraft desired by Coad and Woolman would modify the standard biplane into a stronger structure with a larger engine that would enable the airplane to carry "1,000 pounds of calcium arsenate dust" and permit it to fly with a total maximum weight 5,250 pounds. For the collaboration of Coad, Post, and Woolman on developing a crop duster, see Lewis and Newton, *Delta*, p. 11 and Jones, *Delta Air Lines*, pp. 9–10. While the Huff Daland's dusting operation

centered in the South, its airplanes were still manufactured by its parent company, Huff Daland Airplanes, which changed its name to Keystone Aircraft Corporation after it moved its headquarters from New York to Pennsylvania in 1925. See R. E. G. Davies, *Delta, an Airline and Its Aircraft: The Illustrated History of a Major U.S. Airline* (Miami: Paladwr Press, 1990), p. 9.

10. Lewis and Newton, *Delta*, p. 11 and Jones, *Delta Air Lines*, p. 10. In the pre- and early history of Delta Air Lines, Woolman always acted as the central planner and decision-maker for the company, even though his title sometimes changed during the early years. For example, for legal reasons in 1934, Woolman resigned as director of the company and assumed the title of "general manager." Despite these changes, Woolman is credited with "piloting" the company into success and for always being "chief nerve center of the enterprise." See "C.E. Woolman Is Dead at 76; Chairman of Delta Airlines," *New York Times*, September 12, 1966 and Lewis and Newton, *Delta*, pp. 29, 46.
11. Smoot field was abandoned two years later for the improved Selman Field in 1927. This would be Delta's base of operations until Woolman moved the company to Atlanta in 1941. Lewis and Newton, *Delta*, p. 12.
12. Lewis and Newton, *Delta*, p. 12 and Jones, *Delta Air Lines*, p. 10.
13. Jones, *Delta Air Lines*, p. 13.
14. Lewis and Newton, *Delta*, p. 13
15. Jones, *Delta Air Lines*, p. 10
16. For a depiction of the llama logo, see Davies, *Delta*, p. 9.
17. A more detailed look at Trippe's involvement in the development of Peruvian Airways Incorporated, which eventually became part of the airline Panagra, will be discussed in chapter 4.
18. Harold R. Harris to Irwin Auerbach, C. E. Woolman files, Delta General Offices, Atlanta; quoted in Lewis and Newton, *Delta*, p. 18.
19. Ibid., p. 18.
20. Lewis and Newton, *Delta*, pp. 19–21.
21. Fitzgerald would later become "one of the first women ever elected to the board of directors of an American airline." Lewis and Newton, *Delta*, pp. 17–18, 21.
22. Lewis and Newton, *Delta*, p. 16.
23. Unless otherwise specified, the information about national trends in aviation in the late 1920s was derived from two useful essays that can be found on the *U.S. Centennial of Flight Commission: History of Flight* Web site. See "Airmail: The Airmail Act of 1925 through 1929" and "The Pioneering Years: Commercial Aviation, 1920–1930" http://www.centennialofflight.gov/essay/ (accessed April 13, 2005).
24. Smith, *Airways*, p. 103 and R. E. G. Davies, *A History of the World's Airlines* (London: Oxford University Press, 1964), pp. 40–47.
25. Because night flying was still not practical, trains took over during the evening. Smith, *Airways*, pp. 143–144.
26. Ibid., pp. 145–146.
27. George Thomas Kurian, *Datapedia of the United States 1790–2005* (Lanham, MD: Bernan Press, 2001), pp. 335–336.
28. Lewis and Newton, *Delta*, p. 22. For criticism of the "Lindbergh boom" as being a "myth," see William F. Trimble, "George F. Hahn, Pittsburgh Aviation Corporation, and Pennsylvania Airlines" in *Airline Executives and Federal Regulation: Case Studies in American Enterprise from the Airmail Era to the Dawn of the Jet Age* (Columbus, OH: Ohio University Press, 2000), p. 51.
29. Lewis and Newton, *Delta*, p. 22.

30. Ibid., pp. 21–22.
31. Davies, *A History of the World's Airlines*, p. 53 and Smith, *Airways*, pp. 150–151.
32. See, for instance, a report in the *New York Times* that came out just before Woolman began to set his sights on buying Huff Daland Dusters: "The [passenger] business is growing faster than its most optimistic advocates ever expected." See "Airplanes Now Carry Mail between 31 of Our States," *New York Times*, May 13, 1928.
33. Lewis and Newton, *Delta*, pp. 22–23.
34. Davies, *Delta*, pp. 10, 13.
35. Ibid and Lewis and Newton, *Delta*, p. 23. This would be a crucial first step in establishing Delta's reputation for good customer service. Woolman's focus on customer service was made clear throughout his career at Delta. During many company sales meetings, Woolman said: "Let's put ourselves on the other side of the [ticket] counter. We have a responsibility over and above the price of a ticket." See "C.E. Woolman Is Dead at 76; Chairman of Delta Airlines," *New York Times*, September 12, 1966. Although Travel Air produced good planes, they were certainly not the largest available. The Ford Tri-Motor 5-AT used by National Air Transport in the late 1920s could carry 13 passengers, and the German manufacturer Fokker came out with the 30 passenger F-32 in April 1930. Davies, *A History of the World's Airlines*, p. 54, plate 11B.
36. Davies, *Delta*, p. 10 and Lewis and Newton, *Delta*, pp. 23–24.
37. Lewis and Newton, *Delta*, p. 430 fn.21.
38. Davies, *Delta*, p. 47.
39. Woolman to Harris, April 1930, C. E. Woolman Papers, Delta General Offices, Atlanta Georgia; quoted in Lewis and Newton, *Delta*, p. 24.
40. "Walter Folger Brown: The Postmaster General Who Built the U.S. Airline Industry," in *U.S. Centennial of Flight Commission: History of Flight* http://www.centennialofflight.gov/essay/Commercial_Aviation/Brown/Tran3.htm/ (accessed April 13, 2005).
41. "Walter Folger Brown," *U.S. Centennial of Flight Commission: History of Flight*. In 1942, Delta owned a total of 9 airplanes while American and United owned 74 and 62 respectively. See Davies, *A History of the World Airlines*, p. 136. Delta would emerge as a major airline after a series of mergers and acquisitions after World War II.
42. Smith, *Airways*, p. 168.
43. Lewis and Newton, *Delta*, p. 25.
44. Smith, *Airways*, pp. 167–168.
45. Lewis and Newton, *Delta*, p. 26. Woolman may have received this information from "Seth Barwise, a Texas aviation pioneer." See ibid., p. 431 fn.26.
46. Lewis and Newton, *Delta*, pp. 26–27.
47. Smith, *Airways*, pp. 148–149.
48. Robert J. Serling, *Eagle: The Story of American Airlines* (New York: St. Martin's Press, 1985), pp. 11–12.
49. Lewis and Newton, *Delta*, p. 27.
50. Ibid., p. 27.
51. Ibid., pp. 27–28.
52. Ibid., p. 28.
53. Serling, *Eagle*, p. 33.
54. Christopher Marquis, "Historical Environments, Coordination and Consolidation in the U.S. Banking Industry, 1896–2001." Paper presented at the annual meeting of the Academy of Management in Atlanta, August 2006.
55. Lewis and Newton, *Delta*, pp. 28–29.

56. Woolman to Harris, April 22, 1930, C. E. Woolman files, Delta General Offices, Atlanta; quoted in Lewis and Newton, *Delta*, pp. 30–31. Harris became increasingly interested in working in South America and eventually left Delta to become a vice president of Panagra.
57. Lewis and Newton, *Delta*, p. 31.
58. Ibid., p. 24.
59. See Serling, *Eagle*, p.16.
60. Lewis and Newton, *Delta*, pp. 30–41.
61. David D. Lee, "Herbert Hoover and the Development of Commercial Aviation, 1921–1926," *Business History Review* 58 (1984), p. 79.
62. Lewis and Newton record this exchange of January 11, 1934 between Senator Black and Woolman: "Did the company who got the line [AVCO] have any experience of any kind or character on your line?" Woolman replied, "No one had flown that line but ourselves." A moment later Black asked, "did you sell out because you wanted to sell out?" Woolman replied, "We sold out because it seemed the expedient thing to do." "Why?" Black asked. "Because," Woolman said, "it would be impossible to compete with a line carrying airmail and it was impossible and has proven repeatedly, to make money in carrying passengers alone." Lewis and Newton, *Delta*, p. 42.
63. "Air Mail Expenditures. By Senator Hugo L. Black," *New York Times*, February 18, 1934.
64. Lewis and Newton, *Delta*, p. 42.
65. "Airmail and the Growth of Airlines," *U.S. Centennial of Flight Commission: History of Flight*, http://www.centennialofflight.gov/essay/Government_Role/1930-airmail/POL6.htm.
66. See, for example, "Time to Reconsider," *New York Times*, February 27, 1934 and "Safety for Mail Planes," *New York Times*, March 18, 1934.
67. "Airmail and the Growth of Airlines," *U.S. Centennial of Flight Commission: History of Flight*, http://www.centennialofflight.gov/essay/Government_Role/1930-airmail/POL6.htm.
68. Ibid.
69. For the national impact of this crash, see, for example, "Rockne's Loss Described as a National Loss in Tribute by President Hoover," *New York Times*, April 2, 1931.
70. Judy Rumerman, "Douglas Aircraft Builds the DC-1 and the DC-2," *U.S. Centennial of Flight Commission: History of Flight*, http://www.centennialofflight.gov/essay/Aerospace/Douglas-1930s/Aero28.htm.
71. Davies, *A History of the World's Airlines*, pp. 133–134.
72. Davies, *Delta*, p. 26 and Lewis and Newton, *Delta*, p. 43.
73. Lewis and Newton, *Delta*, Appendix 1.
74. Davies, *Delta*, p. 29 and "Delta History," in *Delta Air Transport Heritage Museum*, http://www.deltamuseum.org/History/main_delta_history.html.
75. Davies, *Delta*, p. 29.
76. The CAB was established by the U.S. Congress with responsibility for "overseeing the federal airways systems, developing airports, issuing airworthiness certificates for aircraft, and certifying airline pilots." Airline historian W. David Lewis notes: "The CAB was one of the most powerful regulatory agencies in American history. Its control of the airline industry was virtually absolute. After 1938, no airline could perform airmail or schedule passenger operations over any route without securing what was called a Certificate of Public Convenience and Necessity (CPCN) from the CAB or its earlier counterpart, the old CAA (Civil Aeronautics Authority). Once service was established on a route, it could not be abandoned without permission. No carrier could engage in commercial aviation without CAB

sanction, and no merger between existing airlines could take place without CAB's consent. All tariffs and charges for the transport of passengers and goods had to receive CAB approval...Airlines could arrange schedules and select aircraft, but do little else without asking Uncle Sam's permission." W. David Lewis, "Introduction—Ambivalent Relationship: Airline Executives and Federal Regulation," in *Airline Executives and Federal Regulation*, ed. W. David Lewis (Columbus, OH: Ohio State University Press, 2000), pp. 12–14.

77. Davies, *Delta*, p. 30. See also Roger Mola, "Economic Regulation of Airlines," in *U.S. Centennial of Flight Commission: History of Flight*, http://www.centennialofflight.gov/essay/Government_Role/Econ_Reg/POL16.htm.
78. There were, for instance, stories concerning an accident of September 1, 1940 in which Senator Ernest Lundeen of Minnesota and 24 other people died in a crash of a Pennsylvania Central Airlines DC-3 in what was, up to that time, the "worst commercial aviation accident." See "Says Fatal Plane Struck at Full Tilt," *New York Times*, September 2, 1940. See also Lewis and Newton, *Delta*, p. 70.
79. Judy Rumerman, "The DC-3" in *U.S. Centennial of Flight Commission: History of Flight*, http://www.centennialofflight.gov/essay/Aerospace/DC-3/Aero29.htm.
80. Lewis and Newton, *Delta*, p. 74.
81. For a detailed discussion of Delta during this late 1930s to 1941, see Lewis and Newton, *Delta*, pp. 73–81.
82. Lewis and Newton, *Delta*, pp. 76–77.
83. Ibid., pp. 77–78.
84. Ibid., pp. 436–7 fn25.
85. Ibid., "Appendix 1."
86. W. David Lewis and Wesley Phillips Newton, "C. E. Woolman," in *Encyclopedia of American Business History and Biography: The Airline Industry,* ed. William M. Leary (New York: Facts on File, 1992), p. 508.
87. Ibid., p. 509.
88. Ibid., p. 512.
89. Ibid., p. 515.
90. Ibid., p. 512.

Part II The Managers

1. Steven Klepper, "The Capabilities of New Firms and the Evolution of the U.S. Automobile Industry," *Industrial and Corporate Change*, Oxford University Press, August 2002, pp. 645–666.
2. "Quotes from C. E. Woolman," The Delta Heritage Museum Web site, http://www.deltamuseum.org/M_Education_DeltaHistory_Facts_FounderQuotes.htm (accessed June 23, 2008).

Chapter 4 Juan Trippe and the Growth of International Air Travel

1. Portions of this chapter have been excerpted from Nitin Nohria, Anthony J. Mayo, and Mark Rennella, "Juan Trippe and Pan American World Airways," HBS Case No. 406–086 (Boston: Harvard Business School Publishing, 2006).
2. Marilyn Bender and Selig Altschul, *The Chosen Instrument: Pan Am, Juan Trippe—The Rise and Fall of an American Entrepreneur* (New York: Simon and Schuster, 1982), pp. 252–253.

3. "The Screen: Warner's 'China Clipper' at Strand Documents Dramatic Story of a Transpacific Flight," *New York Times*, August 12, 1936.
4. For a report of Pan Am's Pacific flight path as of September 1936, see "Many Seek Accommodations on Trans-Pacific Flights," *Wall Street Journal*, September 25, 1936.
5. Bender and Altschul, *The Chosen Instrument*, pp. 196–202. See also "Lindbergh's Chart Flood Area in China," *New York Times*, September 22, 1931.
6. "Lindbergh Pictures Will Aid a Hospital," *New York Times*, October 29, 1931.
7. Bender and Altschul, *The Chosen Instrument*, pp. 201–202.
8. Ibid., p. 202
9. Ibid., p. 223.
10. Ibid., p. 225.
11. Ibid., p. 253.
12. For a discussion of these issues, see Robert H. Van Meter, Jr., "The Washington Conference of 1921–22: A New Look," *Pacific Historical Review* 46 (1977), pp. 603–624.
13. Akira Iriye, *Across the Pacific: An Inner History of American-East Asian Relations* (New York: Harcourt, Brace & World, 1967), p. 175.
14. Bender and Altschul, *The Chosen Instrument*, p. 230.
15. Gerard E. Wheeler, "Isolated Japan: Anglo-American Diplomatic Co-operation, 1927–1926," *Pacific Historical Review* 30 (1961) p. 175.
16. Bender and Altschul, *The Chosen Instrument*, pp. 230–232.
17. For information on Trippe's early connections with the Roosevelt administration, see Bender and Altschul, *The Chosen Instrument*, pp. 216–218. For information on Pan Am being too valuable to U.S. interests to endanger, see Bender and Altschul, *The Chosen Instrument*, pp. 220–223.
18. Bender and Altschul, *The Chosen Instrument*, pp. 233–236.
19. Ibid., pp. 206–208.
20. "Clipper Off Today on Pacific Mail Hop," *New York Times*, November 22, 1935.
21. Bender and Altschul, *The Chosen Instrument*, pp. 253–256.
22. Ibid., pp. 182–184.
23. Ibid., p. 224. See also, "Huge Ocean Planes Ordered by Airline," *New York Times*, December 1, 1932.
24. Richard K. Smith, "The Intercontinental Airliner and the Essence of Airplane Performance, 1929–1939," *Technology and Culture* 24 (1983), p. 434.
25. Bender and Altschul, *The Chosen Instrument*, p. 238.
26. Smith, "The Intercontinental Airliner and the Essence of Airplane Performance," p. 439.
27. Bender and Altschul, *The Chosen Instrument*, p. 264.
28. Smith, "The Intercontinental Airliner and the Essence of Airplane Performance," p. 440 and Bender and Altschul, *The Chosen Instrument*, pp. 264–265.
29. Juan Trippe, Interview of May 25, 1978, quoted in Bender and Altschul, *The Chosen Instrument*, p. 264.
30. As Pan Am explained to its investors in 1949, "The closing of China to American commerce as well as disturbed conditions in Indo-China, Indonesia and Burma adversely affected the volume of trans-Pacific travel." See Pan American World Airways, *Annual Report for 1949*, p. 8.
31. Floyd Norris and Christine Bockelmann, *The New York Times Century of Business* (New York: McGraw-Hill, 2000), p. 112.
32. Kenneth M. Morris, Marc Robinson, and Richard Kroll, *American Dreams, One Hundred Years of Business Ideas and Innovation from The Wall Street Journal* (New York: Lightbulb Press, 1992), p. 101.

33. Robert Daley, *An American Saga – Juan Trippe and His Pan Am Empire* (New York: Random House, 1980), p. 231.
34. Bender and Altschul, *The Chosen Instrument*, pp. 295–296.
35. Ibid., p. 297.
36. Ibid., p. 299.
37. Ibid., p. 301.
38. Ibid., p. 301.
39. Daley, *An American Saga*, p. 335.
40. Ibid., p. 335.
41. Wesley Phillips Newton, "Juan T. Trippe," in *Encyclopedia of American Business History and Biography: The Airline Industry*, ed. William M. Leary (New York: Facts on File, 1992), pp. 472–474.
42. Ibid., p. 473.
43. Ibid., pp. 473–475.
44. Bender and Altschul, *The Chosen Instrument*, p. 365.
45. Ibid.
46. Ibid., p. 477.
47. Daley, *An American Saga*, p. 346.
48. Ibid., pp. 349–350.
49. Ibid., p. 350.
50. Ibid., p. 356.
51. Sources for the sidebar on Donald Nyrop include Donna Corbett, "Donald W. Nyrop: Airline Regulator, Airline Executive," in *Airline Executives and Federal Regulation*, ed. W. David Lewis (Columbus: Ohio State University Press, 2000); "Presidential Error," *Time*, February 21, 1955; Geoff Jones, *Northwest Airlines: The First Eighty Years* (Charleston, SC: Arcadia Publishing, 2005); and John Smiley, "Northwest Airlines Corporation," in *International Directory of Company Histories*, ed. Tina Grant, vol. 74 (Detroit: St. James Press, 2006).
52. Bender and Altschul, *The Chosen Instrument*, p. 516.
53. In 1936, one-way and round-trip fares between San Francisco and Manila were $800 and $1,438, respectively. See "Trans-Pacific Fares," *Wall Street Journal*, October 3, 1936. In 1951, a round-trip ticket between San Francisco and Tokyo was $1,170. The *Wall Street Journal* remarked in 1951 that Pan Am's U.S. to Asia fares had not changed much and were not expected to move in the following year. See "Cheaper Flying," *Wall Street Journal*, October 15, 1951.
54. See, "Cheaper Flying," and "Proposes $96 Hawaii Fare," *New York Times*, August 29, 1944.
55. "Aviation: Airlines Seek to Develop Tourist Trade to Puerto Rico and the Near-by Islands," *New York Times*, September 1, 1946.
56. Daley, *An American Saga*, p. 514.
57. Bender and Altschul, *The Chosen Instrument*, p. 395.
58. "Atlantic Fare Is Slashed to $275," *New York Times*, October 17, 1945.
59. Bender and Altschul, *The Chosen Instrument*, p. 395.
60. Daley, *An American Saga*, p. 514.
61. Ibid., p. 514.
62. "Trippe is Named President-Elect of International Airlines Group," *New York Times*, September 18, 1954.
63. Pan American World Airways, *Annual Report for1954*, p. 1.
64. "Riding High in a Jet," *New York Times*, December 20, 1953.
65. A spokesman for the sole purchaser of the Comet, the British Overseas Airways Corporation (B.O.A.C.), reported that "the four-jet plane is a money maker." See

"Aviation: British Jets," *New York Times*, October 12, 1952. C. R. Smith, speaking at a transportation conference in April 1953, dismissed the jet's utility: "We are, all of us, still intrigued by the glamour of the turbo-jet airplane, but neither we nor you (the customer) can now afford it." See "Turbo-Jet Cost Held Too High for Airlines," *New York Times*, April 30, 1953.

66. At the time, those economies of scale seemed attainable only by aircraft that could carry 180 passengers. See Bender and Altschul, *The Chosen Instrument*, p. 470. A contemporary and popular turbo-prop aircraft, the Lockheed 188 Electra, had a maximum seating capacity of 85. See R. E. G. Davies, *Delta: The Illustrated History of a Major U.S. Airline and the People Who Made It* (Miami: Paladwr Press, 1990), p. 71.
67. "Aviation: Looking Ahead," *New York Times*, July 13, 1952.
68. "The Comet Riddle," *New York Times*, October 24, 1954.
69. For a concise description of the turbo-prop engine, see "Turboprop Engine" in *NASA Glen Technologies Learning Home Page*, http://www.grc.nasa.gov/WWW/K-12/airplane/aturbp.html (accessed October 31, 2005).
70. Daley, *An American Saga*, p. 401.
71. For a short history of Pratt and Whitney, see Judy Rumerman, "Pratt and Whitney," http://www.grc.nasa.gov/WWW/K-12/airplane/aturbp.html (accessed October 31, 2005).
72. Bender and Altschul, *The Chosen Instrument*, p. 474.
73. Ibid., pp. 473–475.
74. Daley, *An American Saga*, pp. 410–412. When Europeans began to petition William Allen for 707 aircraft with the larger J-75 engines, Allen found that he had to conform to Trippe's wishes in the end. Bender and Altschul, *The Chosen Instrument*, p. 475.
75. *L'Enjeu Aerien: Air France* (Paris: Editions France-Empire, 1972), p. 93.
76. Nicolas Neiertz, "Air France: An Elephant in an Evening Suit?" in *Flying the Flag: European Commercial Air Transport since 1945*, ed. Hans-Liudger Dienel and Peter Luth (London: Macmillan Press, 1998), p. 35.
77. Bender and Altschul, *The Chosen Instrument*, p. 500.
78. Ibid.
79. Ibid., p. 503.
80. Harold Evans, *They Made America* (New York: Little, Brown, 2004), p. 299.
81. Daley, *An American Saga*, p. 433.
82. Bender and Altschul, *The Chosen Instrument*, p. 507.
83. Ibid., p. 507.
84. Ibid., p. 510.
85. Pan American World Airways, *Annual Report for 1968*.
86. "Juan Trippe, 81, Dies; U.S. Aviation Pioneer," *New York Times*, April 4, 1981.
87. Bender and Altschul, *The Chosen Instrument*, p. 517.
88. Evans, *They Made America*, p. 300.
89. Richard Branson, "Pilot of the Jet Age—Juan Trippe," *Time*, December 7, 1998.

Chapter 5 C. R. Smith and American Airlines

1. Portions of this chapter are drawn from Anthony J. Mayo and Laura G. Singleton, "C.R. Smith and the Birth of American Airlines," HBS Case No. 406–082 (Boston: Harvard Business School Publishing, 2006) and Anthony J. Mayo, Nitin Nohria, and Laura G. Singleton, *Paths to Power: How Insiders and Outsiders Shaped American Business Leadership* (Boston: Harvard Business School Press, 2006), pp. 48–53.
2. Robert J. Serling, *Eagle: The Story of American Airlines* (New York: St. Martin's Press, 1985), p. 86.

3. Ibid., p. 86.
4. Ibid., p. 78.
5. Jack Alexander, "Just Call Me C.R.," *Saturday Evening Post*, February 1, 1921, p. 72.
6. Ibid., p. 10.
7. Ibid., p. 10.
8. Ibid., p. 69.
9. John N. Ingham, *Biographical Dictionary of American Business Leaders* (Westport, CT: Greenwood Press, 1983), p. 1316.
10. Serling, *Eagle*, pp. 45, 47.
11. Alexander, "Just Call Me C.R.," p. 69.
12. Serling, *Eagle*, p. 47
13. Smith compiled names and addresses of stockholders from Texas state corporate listings and sold them to investment firms, then expanded to compiling lists of new parents from state birth records and selling these to companies offering products for families with young children. The lists were particularly useful for parents' magazines, infant food companies, and manufacturers of nursery and baby buggy equipment. See Alexander, "Just Call Me C.R.," *Saturday Evening Post*, p. 70 and Serling, *Eagle*, p. 47.
14. Serling, *Eagle*, p. 47.
15. Ingham, *Biographical Dictionary of American Business Leaders*, p. 1317.
16. Serling, *Eagle*, p. 45.
17. Ibid.
18. For a concise description of the consolidation of the airlines between 1925 and 1930, see R. E. G. Davies, *A History of the World's Airlines* (London: Oxford University Press, 1964), p. 46.
19. Serling, *Eagle*, p. 10 and Charles J. Kelly, Jr., *The Sky's the Limit: The History of the Airlines* (New York: Coward-McCann, 1963), p. 213.
20. Serling, *Eagle*, p. 11.
21. C. R. Smith, *"A.A." American Airlines—since 1926* (New York: Newcomen Society of North America, 1954), p. 11.
22. Serling, *Eagle*, p. 16.
23. Ibid.
24. For details on the battle for corporate control of American Airways, see "Airline in the Black," *Fortune* 19 (February 1939), p. 115.
25. Serling, *Eagle*, pp. 43–44.
26. "Airline in the Black," p. 115.
27. Serling, *Eagle*, p. 55.
28. Ibid., pp. 55–56.
29. Serling describes the reactions of passengers to the Condor. They "loved the Condor's roomy comfort. In American's configuration, the big biplane carried eighteen passengers by day and could accommodate fourteen in the surprisingly spacious berths. Compared with the noisy, rattling Fords and cramped Stinsons and Vultees [the prevailing aircraft], the Condor interior was palatial and compensated for the airliner's abysmal performance. The top speed was only 145 miles per hour, and that was achieved on extremely rare occasions." See Serling, *Eagle*, p. 56.
30. Paul P. Willis, *Your Future Is in the Air: The Story of How American Airlines made People Air-Travel Conscious* (New York: Prentice Hall, 1940), pp. 59–62 and Serling, *Eagle*, p. 56.
31. Serling, *Eagle*, p. 57 and Roger E. Bilstein, "C.R. Smith," in *Encyclopedia of American Business History and Biography: The Airline Industry*, ed. William M. Leary (New York: Facts on File, 1992), p. 437.

32. Serling quotes Littlewood on the Condor: "It was a dangerous airplane. The frame was welded steel, but the fuselage and wings were cloth-covered and there was no firewall between the engine and the baffle. Two [Condors] burned up on the ground at Los Angeles. It was phenomenal luck that a Condor never caught fire in the air." See Serling, *Eagle*, p. 57.
33. The information about Braniff was derived from Roger E. Bilstein, "Thomas E. Braniff," in *Encyclopedia of American Business History and Biography: The Airline Industry*, ed. William M. Leary (New York: Facts on File, 1992), pp. 73–80 and The Braniff Pages, "1936—The Tom Braniff Years," at http://www.braniffpages.com/1930/1930.html (accessed August 21, 2007).
34. See Serling, *Eagle*, pp. 69–70.
35. Ibid., p. 72.
36. Felix Bruner, "Air Mail Cost Lowest Ever in Bids from Rebuilt Lines," *The Washington Post*, April 21, 1934 and Alexander, "Just Call Me C.R.," *Saturday Evening Post*, p. 9.
37. "Airline in the Black," p. 116.
38. Ibid.
39. Kelly, Jr., *The Sky's The Limit*, p. 213.
40. Smith, *"A.A.,"* p. 15.
41. Serling, *Eagle*, pp. 81–82.
42. Ibid., p. 87.
43. Ibid., p. 89.
44. Ibid., p. 90.
45. Ibid., pp. 90–91.
46. Ibid., p. 91.
47. Ibid., p. 101.
48. James. D. Matthews, "The DC-3 Opens a New Era of Commercial Air Travel," in *Great Events from History II: Business and Commerce*, ed. Frank N. Magill (Pasadena, CA: Salem Press, 1994), p. 754.
49. Curt Schleier, "He Kept Carrier Flying High," *Investor's Business Daily*, July 9, 2002.
50. Smith, *"A.A.,"* p. 15.
51. Serling, *Eagle*, p. 109.
52. Airmail and passenger revenue figures for American Airlines can be found in the 1935 and 1936 Annual Reports for American Airlines. American Airlines, *1935 Annual Report* (Chicago: American Airlines, 1935), p. 2 and American Airlines, *1936 Annual Report* (Chicago: American Airlines, 1936), p. 2.
53. "Little-Known Facts," Air Transport Association of America, January 1, 1937.
54. Serling, *Eagle*, p. 110.
55. Ibid., p. 101.
56. Willis, *Your Future Is in the Air*, p. 74.
57. Patterson's main reaction to the accidents was to hire Major R. W. Schroeder, a "veteran of World War I Army Signal Corps" who had also "served with the Bureau of Air Commerce in charge of air safety." According to one historian of United, Schroeder "lived and dreamed air safety.... If any man could lick the bugaboo of airliner crashes, Schroeder could do it." See Frank J. Taylor, *High Horizons: Daredevil Flying Postmen to Modern Magic—The United Air Lines Story* (New York: McGraw-Hill, 1951), p. 107.
58. Willis, *Your Future Is in the Air*, pp. 79–81.
59. Ibid., p. 81.
60. Roger E. Bilstein, "Air Travel and the Traveling Public," in *From Airships to Airbus: The History of Civil and Commercial Aviation*, eds. William M. Leary and William F. Trimble, 2 vols. (Washington, DC: Smithsonian Books, 1995), 2: p. 95.

61. Serling, *Eagle*, p. 82.
62. American Airlines, *1937 Annual Report* (Chicago: American Airlines, 1937), p. 4.
63. Serling, *Eagle*, p. 110.
64. Bilstein, "Air Travel and the Traveling Public," pp. 95–96.
65. Serling, *Eagle*, p. 99.
66. "American Airlines Had $213,262 Profit," *New York Times*, March 16, 1939; "TWA Shows Loss of $773,263 for Year Ended December 31, 1938," *Wall Street Journal*, February 23, 1939; and "United Air Lines Reports $997,221 Loss for 1938," *Wall Street Journal*, March 17, 1939.
67. "American Airlines Had $213,262 Profit."
68. Serling, *Eagle*, p. 152 and "American Airlines Raises Executive," *New York Times*, March 5, 1942.
69. American Airlines had suffered a fatal accident during the dark days of December 1935 and January 1936 when one of its aircraft plowed into an Arkansas swamp killing 17 persons on January 14, 1936. See Alexander, "Just Call Me C.R.," *Saturday Evening Post* and Serling, *Eagle*, p. 152.
70. Alexander, "Just Call Me C.R.," *Saturday Evening Post*, p. 9 and Serling, *Eagle*, p. 152.
71. Serling, *Eagle*, p. 167.
72. Ibid., pp. 179, 183.
73. Ibid., p. 183.
74. Roger E. Bilstein, "C.R. Smith: An American Original," in *Airline Executives and Federal Regulation*, ed. W. David Lewis (Columbus, OH: Ohio State University Press, 2000), p. 90.
75. Kelly, Jr., *The Sky's the Limit*, p. 223.
76. Bilstein, "C.R. Smith: An American Original," p. 91.
77. Ibid., p. 96
78. Serling, *Eagle*, p. 192.
79. Bilstein, "C.R. Smith: An American Original," p. 96.
80. Serling, *Eagle*, p. 196.
81. Ibid., p. 196.
82. Bilstein, "C.R. Smith: An American Original" and "Jets Across the U.S.," *Time*, November 17, 1958, p. 88.
83. Smith, *"A.A.,"* p. 18.
84. Serling, *Eagle*, pp. 197–199.
85. The design of the DC-6 was a joint effort between Douglas, United and American. With 4 engines, it had the capability to go 300 miles per hour. The DC-6 was also pressurized that enabled the aircraft to cruise at higher altitudes, which, in turn, provided more passenger comfort while circumventing low-level weather-related delays. See Bilstein, "C.R. Smith: An American Original" and Serling, *Eagle*, pp. 196–197.
86. Serling, *Eagle*, p. 217.
87. American Airlines, *1947 Annual Report* (New York: American Airlines, 1947) and American Airlines, *1948 Annual Report* (New York: American Airlines, 1948).
88. Serling, *Eagle*, p. 217.
89. Smith, *"A.A.,"* p. 18.
90. Bilstein, "C. R. Smith: An American Original," p. 98.
91. C. R. Smith, "What We Need Is a Good Three-Cent Airline," *Saturday Evening Post*, October 20, 1945, p. 12.
92. Serling, *Eagle*, p. 236.
93. Ibid., p. 236.

94. American Airlines, *1950 Annual Report* (New York: American Airlines, 1950) and American Airlines, *1960 Annual Report* (New York: American Airlines, 1960).
95. According to Serling: "The CAB's initial reaction to the scheduled carriers' wave of coach service was unfavorable. It even cautioned the airlines that it did not intend 'to permit a general debasement of the existing passenger fare level.' But C. R. sensed that this mood wasn't going to last, and his crystal ball was crystal clear." See Serling, *Eagle*, p. 259.
96. Bilstein, "C. R. Smith: An American Original," p. 102.
97. American Airlines, "American Airlines History," American Airlines Web site, http://www.amrcorp.com/history.htm (accessed April 26, 2002).
98. Ibid.
99. Serling, *Eagle*, p. 348. See also Neil Ulman, "Versatile Computers," *Wall Street Journal*, May 27, 1963. For a detailed description of the workings of SABRE, see "Computers Used to Unsnarl Airline-Ticket Snags," *New York Times*, September 16, 1964.
100. Bilstein, "C.R. Smith: An American Original," p. 93.
101. Ibid., p. 94.
102. Ibid., p. 95.
103. "Riding High in a New Era," *Nation's Business* 54 no. 3 (March 1966).
104. Douglas Karsner, " 'Now Hawaii Is only Hours Away!': The Airlines Alter Tourism," *Essays in Economic and Business History* 17 (1999), pp. 182–183. Louisa Comstock, "Put Your Two Weeks on Wings," *Better Homes and Gardens* 26 (September 1947), pp. 214–216, quoted in Karsner, "The Airlines Alter Tourism," p. 184.
105. H. Peter Gray, *International Travel—International Tourism* (Lexington, MA: D.C. Heath, 1970), pp. 170–171, cited in Karsner, p. 190.
106. Bilstein, "C. R. Smith: An American Original," p. 107.
107. Statistics comparing the revenue passenger miles of various airlines from 1960 to 1961 taken from House Committee of the Judiciary, *Proposed Merger of Eastern Air Lines and American Airlines* (87th Congress, 2nd Session, March 23, 1962), p. 8.
108. Bilstein, "C.R. Smith: An American Original," pp. 108–112.
109. "Jets Across the U.S.," p. 88.
110. Ibid., p. 88.
111. Serling, *Eagle*, p. 280.
112. "Jets Across the U.S.," p. 82.
113. Bilstein, "C. R. Smith: An American Original," p. 116.
114. Serling, *Eagle*, p. 311.
115. Dan Reed, *The American Eagle: The Ascent of Bob Crandall and American Airlines* (New York: St. Martin's Press, 1993), p. 31.
116. Serling, *Eagle*, p. 419.

Chapter 6 William "Pat" Patterson and United Air Lines

1. Selig Altschul, *United Air Lines, Inc.: A Reappraisal* (Chicago, IL: United Air Lines, 1953), p. 30.
2. Altschul, *United Air Lines, Inc.*, p. 7.
3. Patterson quoted in Frank J. Taylor, *"Pat" Patterson* (Menlo Park, CA: Lane Magazine & Book Company, 1967), p. 147.
4. Patterson quoted in Taylor, *"Pat" Patterson*, p. 121.

5. Altschul, *United Air Lines, Inc.*, p. 30.
6. Taylor, *"Pat" Patterson*, p. 78.
7. "William Allan Patterson," *Dictionary of American Biography, Supplement 10: 1976–1980* (New York: Charles Scribner's, 1995).
8. F. Robert van der Linden, "William Allan Patterson," in *Encyclopedia of American Business History and Biography: The Airline Industry,* ed. William M. Leary (New York: Facts on File, 1992), p. 358.
9. Taylor, *"Pat" Patterson*, pp. 18–20.
10. Ibid., p. 15.
11. Linden, "William Allan Patterson," p. 352.
12. Taylor, *"Pat" Patterson*, pp. 16–17.
13. Linden, "William Allan Patterson," p. 352.
14. Taylor, *"Pat" Patterson*, p. 15.
15. Ibid., p. 24.
16. Ibid., p. 11 and Linden, "William Allan Patterson," pp. 352–353.
17. Linden, "William Allan Patterson," pp. 352–354; Taylor, *"Pat" Patterson*, pp. 25–29; and "Boeing: History," Boeing Web site at http://www.boeing.com/history/index.html (accessed August 27, 2007).
18. "Boeing: History," Boeing Web site at http://www.boeing.com/history/index.html (accessed August 27, 2007); R. E. G. Davies, *A History of the World's Airlines* (London: Oxford University Press, 1964), p. 46; and *Corporate and Legal History of United Air Lines* (Chicago: Twentieth Century Press, 1953), pp. 3–4.
19. "P. G. Johnson Dies; Aircraft Official," *New York Times,* September 15, 1944 and Linden, "William Allan Patterson," pp. 354–355.
20. Frank J. Taylor, *High Horizons* (New York: McGraw-Hill Book, 1951), p. 58.
21. Linden, "William Allan Patterson," p. 355 and Davies, *A History of the World's Airlines*, p. 46.
22. Linden, "William Allan Patterson," p. 355.
23. Taylor, *High Horizons*, p. 70.
24. Ibid., pp. 70–71.
25. Robert E. Johnson, *Airway One* (Chicago: Lakeside Press, 1974), p. 45.
26. Reginald M. Cleveland, "Contact," *New York Times,* February 28, 1932. In this article, Cleveland also comments on Eastern Air Transport's decision to follow Boeing's lead by introducing air "hostesses" to their airline service.
27. Taylor, *"Pat" Patterson*, pp. 35–39 and Johnson, *Airway One*, pp. 44–45.
28. Taylor, *"Pat" Patterson*, p. 80.
29. Ibid., pp. 14–15.
30. Taylor, *High Horizons*, p. 106 and Linden, "William Allan Patterson," p. 357.
31. Taylor, *"Pat" Patterson*, pp. 59–60.
32. Ibid., pp. 60–62.
33. The writer goes on to describe the advantages of the new flight kitchens: "No longer are large ham and cheese sandwiches accompanied by leathery potato chips passed out in containers to scared passengers. Instead, a hot four-course meal is served on portable tables which fit across the arms of the seats in the planes." See "Home Comforts for Air Travel Her Objective," *Washington Post*, January 14, 1938.
34. Robert J. Serling, *Maverick: The Story of Robert Six and Continental Airlines* (New York: Doubleday, 1974), pp. 92–93.
35. This diagram is a modified version of the "Three Rings of Perceived Value" developed by Achieve International. See Jim Clemmer (with Barry Sheehy and Achieve International/Zenger-Miller Associates), *Firing on All Cylinders: The Service/*

Quality System for High-Powered Corporate Performance (Homewood, IL: Business One Irwin, 1992), p. 31.

36. Linden, "William Allan Patterson," p. 356.
37. Taylor, *High Horizons*, p. 97.
38. *Corporate and Legal*, pp. 130–131.
39. Ibid., p. 131.
40. Ibid., p. 132.
41. For American Airlines, see "Airline in the Black," *Fortune* 19 (February 1939), p.116. For United, see *Corporate and Legal*, p. 131.
42. "Airmail and the Growth of the Airlines," U.S. Centennial of Flight Commission Web site at http://www.centennialofflight.gov/essay/Government_Role/1930-airmail/POL6.htm (accessed August 23, 2007).
43. Taylor, *High Horizons*, p. 101.
44. Ibid., pp. 102–103 and *Corporate and Legal*, p. 395.
45. "P. G. Johnson Dies; Aircraft Official," *New York Times*, September 15, 1944.
46. Taylor, *"Pat" Patterson*, p. 51.
47. Taylor, *High Horizons*, p. 104.
48. While United did succeed in buying a route to Denver from Wyoming Air Services, the offer to buy a far larger route connecting Milwaukee to Washington via Cleveland and Pittsburgh was denied by the Interstate Commerce Commission. See Taylor, *High Horizons*, p. 104.
49. Taylor, *"Pat" Patterson*, p. 50 and *Corporate and Legal*, p. 421.
50. Taylor, *High Horizons*, p. 103.
51. Judy Rumerman, "Boeing's Metal Monoplanes of the 1930s," at U.S. Centennial of Flight Commission Web site at http://www.centennialofflight.gov/essay/Aerospace/Boeing's_Metal_Planes/Aero18.htm (accessed August 23, 2007).
52. R. E. G. Davies, *Delta, an Airline and Its Aircraft: The Illustrated History of a Major U.S. Airline* (Miami: Paladwr Press, 1990), p. 31.
53. Roger E. Bilstein, *The American Aerospace Industry: From Workshop to Global Enterprise* (New York: Twayne, 1996), p. 57.
54. R. E. G. Davies, *Airlines of the United States Since 1914* (London: Putnam, 1972), p. 186.
55. Robert J. Serling, *Eagle: The Story of American Airlines* (New York: St. Martin's Press, 1985), p. 90 and Johnson, *Airway One*, pp. 26–27.
56. Judy Rumerman, "The Douglas DC-3," at U.S. Centennial of Flight Commission Web site at http://www.centennialofflight.gov/essay/Aerospace/DC-3/Aero29.htm (accessed August 23, 2007).
57. Johnson, *Airway One*, p. 27.
58. Taylor, *High Horizons*, p. 110. Johnson writes, "One advertisement depicted a 247 speeding through starlit midnight skies with a brash headline: 'Swift and Silent as the Night.'" Johnson, *Airway One*, p. 27.
59. Taylor, *High Horizons*, pp. 110–111.
60. *Corporate and Legal*, p. 371.
61. The information on Jack Frye and TWA was derived from the following sources: Robert J. Serling, *Howard Hughes' Airline: An Informal History of TWA* (New York: St. Martin's Press, 1983); Patricia A. Michaelis, "Jack Frye," in *Encyclopedia of American Business History and Biography: The Airline Industry*, ed. William M. Leary (New York: Facts on File, 1992); "Airliner Crosses the Country in 7 Hours, Setting a Record," *New York Times*, April 18, 1944; "The Nation," *New York Times*, April 23, 1944; "6 Hours 58 Minutes," *New York Times*, April 19, 1944; Robert W. Rummel, *Howard Hughes and TWA* (Washington: Smithsonian Institution Press, 1991); "Plane Crash Laid to

Electric Fire," *New York Times*, August 1, 1946; and "TWA Reports Net Loss of $3,235,491 During September Quarter," *Wall Street Journal*, November 27, 1946.
62. *Corporate and Legal*, pp. 371–372.
63. Ibid., pp. 371–372.
64. Ibid., p. 373.
65. Taylor, *"Pat" Patterson*, p. 89.
66. Johnson, *Airway One*, p. 52 and *Corporate and Legal*, pp. 360–361.
67. "Not until 1946 did the DC-4 see commercial service, when United acquired 25 army surplus planes and operated them until Douglas could build and deliver a larger, faster, pressurized aircraft in 1947, the famous DC-6." Johnson, *Airway One*, p. 52.
68. "Order Army to Run Air Lines," *Chicago Daily Tribune*, May 15, 1942.
69. Taylor, *"Pat" Patterson*, pp. 101–102.
70. *Corporate and Legal*, p. 390.
71. "Task of Air Lines Outlined by Army," *Los Angeles Times*, July 22, 1942. United also helped with pilot training at its Boeing School of Aeronautics in Oakland, CA and assisted in making last-minute modifications to the Boeing B-17 bombers at United's maintenance base in Cheyenne, Wyoming. See Johnson, *Airway One*, p. 59.
72. *Corporate and Legal*, pp. 391–393, 395.
73. Linden, "William Allan Patterson," p. 359.
74. Ibid., p. 359.
75. Ibid., p. 358.
76. Altschul, *United Air Lines, Inc.*, p. 50.
77. Ibid., p. 25.
78. Davies, *Delta*, p. 55.
79. Altschul, *United Air Lines, Inc.*, p. 52.
80. Linden, "William Allan Patterson," p. 359 and *Corporate and Legal*, p. 480.
81. Altschul, *United Air Lines, Inc.*, pp. 7, 10, 39–40.
82. "United Air Lines Timeline," United Air Lines Web site at http://www.united.com/page/article/0,6722,2734,00.html (accessed August 23, 2007).
83. Altschul, *United Air Lines, Inc.*, p. 32 and Taylor, *High Horizons*, pp. 183–184.
84. Taylor, *High Horizons*, pp. 183–184.
85. Altschul, *United Air Lines, Inc.*, pp. 33–34.
86. Johnson, *Airway One*, pp. 85–86.
87. Linden, "William Allan Patterson," p. 360.
88. Davies, *Delta*, p. 63; Peter W. Brooks, *The Modern Airliner: Its Origins and Development* (London: Putnam, 1961), pp. 105–106; and Linden, "William Allan Patterson," p. 360.
89. Linden, "William Allan Patterson," p. 362.
90. Patterson quoted in Taylor, *"Pat" Patterson*, p. 133.
91. Taylor, *"Pat" Patterson*, p. 23.

Part III The Leaders

1. In his HBR article "The Theory of Business," Peter F. Drucker discusses how companies that were once dominant become obsolete by not recognizing or affirming changes in the competitive landscape. He writes: "The story is a familiar one: a company that was a superstar only yesterday finds itself stagnating and frustrated, in trouble and, often, in a seemingly unmanageable crisis. . . . The assumptions on

which the organization has been built and is being run no longer fit reality." Peter F. Drucker, "The Theory of the Business," *Harvard Business Review*, September–October 1994, p. 95.

2. Anita McGahan and Julia Kou, "The U.S. Airline Industry in 1995," HBS Case No. 795–113 (Boston: Harvard Business School Publishing, 1995), p. 3. See also Donald R. Whitnah, "Airline Deregulation Act of 1978," in *Encyclopedia of American Business History and Biography: The Airline Industry,* ed. William M. Leary (New York: Facts on File, 1992), p. 15.
3. While American Airlines has had 5 CEOs since 1968, one of their CEOs, Robert Crandall, served for 18 years (1980–1998), and he is largely credited for the success of the airline after deregulation. No other CEO of a major airline served as long as Crandall. The next longest tenure belonged to Robert Allen of Delta who served as CEO of that airline for 10 years from 1987 to 1997.
4. Pankaj Ghemawat and Nancy Donohue, "The U.S. Airline Industry, 1978–1988 (A)," HBS Case No. 390–025 (Boston: Harvard Business School Publishing, 1989), p. 2.
5. McGahan and Kou, "The U.S. Airline Industry in 1995," p. 2.
6. Ibid., p. 3.
7. Whitnah, "Airline Deregulation Act of 1978," p. 16. Ghemawat and Donohue note that "because input costs could be passed on to customers during regulation, airlines generally acceded to union demands and granted regular pay hikes." Ghemawat and Donohue, "The U.S. Airline Industry, 1978–1988 (A)," p. 5.
8. McGahan and Kou, "The U.S. Airline Industry in 1995," p. 4.
9. Whitnah, "Airline Deregulation Act of 1978," p. 16.
10. Ghemawat and Donohue, "The U.S. Airline Industry, 1978–1988 (A)," p. 4; George Thomas Kurian, *Datapedia of the United States 1790–2005* (Lanham, MD: Bernan Press, 2001), p. 335; and McGahan and Kou, "The U.S. Airline Industry in 1995," p. 4.
11. Ghemawat and Donohue, "The U.S. Airline Industry, 1978–1988 (A)," p. 13.
12. McGahan and Kou, "The U.S. Airline Industry in 1995," p. 4.

Chapter 7 Herb Kelleher at Southwest Airlines

1. Bill Freeman, "Kelleher KO'D in Dallas Match Leaves Lawyers in the Lurch," *Air Transport World*, May 1, 1992 and Dee Gil, "Arm Wrasslin' Executives Stage Company PR Spectacle," *Houston Chronicle*, March 21, 1992.
2. Kelleher quoted in Diane Reischel, "Herb Kelleher: The Head of Southwest Airlines Is a Brainy, Free-wheeling Competitor, a Soft-hearted Irishman Whose Party Antics and Hectic Schedule Set the Tone for His Company's Unconventional Spirit," *Dallas Morning News*, April 7, 1985.
3. Kelleher quoted in Reischel, "Herb Kelleher," *Dallas Morning News*, April 7, 1985.
4. According to former Southwest vice president, "Herb operates like a 24-hour 7-Eleven. He lives, eats, sleeps and drinks Southwest Airlines." Reischel, "Herb Kelleher," *Dallas Morning News*, April 7, 1985.
5. Reischel, "Herb Kelleher," *Dallas Morning News*, April 7, 1985.
6. Ibid.
7. Kevin Kelly, "Southwest Airlines: Flying High with 'Uncle Herb,' " *Business Week*, July 3, 1989, p. 53.
8. Edward O. Welles, "Captain Marvel (Herb Kelleher; Southwest Airlines; Entrepreneur of the Year Awards)," *Inc.*, January 1, 1992.

9. Gary Kissel, *Poor Sailors' Airline: A History of Pacific Southwest Airlines* (McLean, VA: Paladwr Press, 2002), p. 172.
10. Nicholas C. Chriss, "Bottles of Scotch, Ice Chests Offered," *Los Angeles Times*, February 12, 1973. See also idem, "Southwest Airlines: Where's LUV Going Without Muse?" *Los Angeles Times*, March 31, 1978.
11. Harold E. Shenton, "David against Goliaths," *The Aeroplane*, February 3, 1966, p. 46.
12. Meyers K. Jacobson, " 'Catch Our Smile': A History of Pacific Southwest Airlines," *American Aviation Historical Society Journal*, Fall 2000, pp. 177–178.
13. PSA charged $9.95 on a flight from Burbank to Oakland while the interstate competition charged $25. John Wegg, "PSA: The Smile Has Gone," *Airliners* 1 (1998), p. 4.
14. Shenton, "David against Goliaths," p. 46.
15. Jacobson, "Catch our Smile," p. 180. This practice of open seating would also be closely associated with Southwest Airlines.
16. Jacobson, "Catch Our Smile," p. 181. The semiofficial history of Southwest Airlines features within its first pages a picture of a Southwest stewardess perched in the overhead compartment handing a coke and peanuts to a delighted passenger. The caption reads: "Southwest flight attendants make flying fun with comical PA announcements and preflight tricks like popping out of overhead bins." See Kevin Freiburg and Jackie Freiburg, *Nuts! Southwest Airlines' Crazy Recipe for Business and Personal Success* (Austin, TX: Bard Press, 1996), p. 6.
17. George Hopkins, "The Texas Airline War," *Washington Monthly*, March 1976, p. 13.
18. In the case of stewardesses, PSA's history did not just repeat itself at Southwest. PSA's introduction of scantily dressed stewardesses added a little something extra to an airline with an already well-established identity and reputation in California. For Southwest, beguiling women serving peanuts and drinks were far more crucial to the airline. With almost no money for marketing in its first years, Southwest depended on word-of-mouth and other forms of friendly reporting on the airline. The stewardesses did make a big impression, appearing on the cover of *Espire* in 1974 and even as far afield as a newspaper in Budapest. Herb Kelleher explained in 1989: "If it hadn't been for the power of that campaign, it's possible Southwest might not have survived its early years." Suzanne Loeffelhoz, "The Love Line: How the Ninth-Largest Airline Continues to Tap-Dance around the Industry Giants," *FW*, March 21, 1989.
19. Reischel, "Herb Kelleher," *Dallas Morning News*, April 7, 1985.
20. Ibid.
21. Ibid.
22. After Southwest cleared its first major legal hurdles, one commentator on Texas business stated: "It's no wonder that Southwest Airlines thinks it can offer cheaper service than anyone else. It won't use aviation fuel—just political power." Hopkins, "The Texas Airline War," p. 14.
23. Freiburg and Freiburg, *Nuts!*, p. 16.
24. Hopkins, "The Texas Airline War," p. 15.
25. Freiburg and Freiburg, *Nuts!*, pp. 17–18.
26. Reischel, "Herb Kelleher," *Dallas Morning News*, April 7, 1985.
27. Kelleher quoted in Freiberg and Freiberg, *Nuts!*, p. 18.
28. With a surplus production, "Boeing offered to sell the planes for $4 million each (comparable new aircraft would have cost $5 million apiece at that time) and finance 90% of the deal." Freiberg and Freiberg, *Nuts!*, p. 19.
29. Freiberg and Freiberg, *Nuts!*, pp. 22–23.

30. Sveinn Vidar Gudmundsson, *Flying too Close to the Sun: The Success and Failure of the New-entrant Airlines* (Aldershot, England: Ashgate, 1998), pp. 12–14.
31. Jacobson, "Catch Our Smile," p. 189.
32. Ibid., p. 187.
33. Ellen Hume, "California May Lose Bargain Air Fares," *Los Angeles Times*, March 16, 1978.
34. Gudmundsson, *Flying too Close to the Sun*, p. 13.
35. See Freiberg and Freiberg, *Nuts!*, pp. 23–26. For Kelleher's relationship with Bob Packwood, see Allen R. Myerson, "Air Herb," *New York Times*, November 9, 1997.
36. The writers of *Nuts!*, citing Kelleher, write that "Southwest Airlines is obsessed with keeping costs low to maximize profitability instead of being concerned with increasing market share." Freiburg and Freiburg, *Nuts!*, p. 49.
37. Allen R. Myerson, "Air Herb," *New York Times*, November 9, 1997.
38. Lamar Muse, *Southwest Passage: The Inside Story of Southwest Airlines' Formative Years* (Austin, TX: Eakin Press, 2002), p. 78.
39. Air Transport Association, "Smart Skies; a Blueprint for the Future: 2006 Economic Report," available at www.ailrines.org.
40. Muse, may have, in fact, been forced out of Southwest because of his attempt to place the company too tightly under his own control. Most sources point to a feud between Rollin King and Muse as the cause behind Muse's departure. But the *Los Angeles Times* added a unique piece of information: "Another source said the straw that broke the camel's back was the fact that Muse had promoted his son, Michael, to be the chief financial officer—setting up a sort of 'Muse Dynasty'—that forced Muse's resignation. The son also resigned. 'Muse was in the old mold of Eddie Rickenbacker . . . or Bob Six at Continental—a one man show or no show at all. Muse didn't need the money, he just needed the power,' said a source from another airline." See Nicholas C. Chriss, "Southwest Airlines: Where's LUV Going without Muse?" *Los Angeles Times*, March 31, 1978. Muse denies the accusation of nepotism. See Muse, *Southwest Passage*, p. 190.
41. Chriss, "Where's LUV Going without Muse," *Los Angeles Times*, March 31, 1978 and "Southwest Air's Chief Says He Left Position in Feud with Holder," *Wall Street Journal*, April 3, 1978.
42. "Southwest Airlines Swaps Idea for Exec," *Chicago Tribune*, August 2, 1978.
43. Kelleher and Putnam co-signed Southwest's Annual Reports. See, for instance, Southwest Airlines, Inc., *Southwest Airlines Annual Report for 1979*. Note that Kelleher's picture and signature precede those of Putnam.
44. Freiburg and Freiburg, *Nuts!*, p. 79.
45. James J. Heskett and Roger H. Hallowell, "Southwest Airlines: 1993 (A)," HBS Case No. 694–023 (Boston: Harvard Business School Publishing, 1993), p. 2.
46. For a description of the development of the hub and spoke system, see Andrew R.Goetz and Christopher J. Sutton, "The Geography of Deregulation in the U.S. Airline Industry," *Annals of the Association of American Geographers* 87 (1997), pp. 238–263 and Stephen D. Solomon, "The Bully of the Skies Cries Uncle: Why Robert Crandall Is Threatening to Pull American Airlines out of the Airline Business," *New York Times Magazine*, September 5, 1993, p. 34.
47. Drew Whitelegg, "Keeping Their Eyes on the Skies," *Journal of Transport History*, 21 (2000), p. 79.
48. Delta also had the distinction of being the first to board one million passengers in one city in one month. It did so in Atlanta in 1979. "Delta Firsts," The Delta

Heritage Museum Web site, http://www.deltamuseum.org/M_Education_DeltaHistory_Facts_Firsts.htm (accessed June 23, 2008).

49. Alfred E. Kahn, "Surprises of Airline Deregulation," *American Economic Review*, 78 (1988), pp. 316, 320 and Bill Sing, "Frills over Fares," *Los Angeles Times*, March 30, 1983.
50. John Newhouse, *The Sporty Game* (New York: Knopf, 1982), p. 84.
51. Southwest Airlines, Inc., *Southwest Airlines Annual Report for 1979*. In 1991, an airline analyst from Goldman Sachs confirmed that Kelleher had remained on course since 1978: "It's remarkable that with all the glitter of deregulation, Southwest has the most disciplined growth of all the entrants—10 to 20 percent a year." Peter C. T. Elsworth, "Southwest Air's New Push West," *New York Times*, June 16, 1991.
52. Southwest Airlines, Inc., *Southwest Airlines Annual Report for 1981*.
53. Freiburg and Freiburg, *Nuts!*, p. 74.
54. Albert R. Karr, "Airline Deregulation after Braniff's Fall," *Wall Street Journal*, June 14, 1982.
55. For a detailed description of Braniff's woes in 1982, see G. Christian Hill, "Braniff Seen Facing Ultimate Liquidation with Other Carriers, Rivals Sole Winners," *Wall Street Journal*, May 17, 1982.
56. "The Chairman of the Board Looks Back," *Fortune*, May 28, 2001.
57. In 1991, the *New York Times* reported that Southwest "hires only one person out of 10 who apply, and those who join the company are made to feel like part of a family.... 'There's no question that the glory of Southwest Airlines is its productivity,' said Edward J. Starkman, airline analyst with PaineWebber. 'The work force is dedicated to the company. They're Moonies basically. That's the way they operate.'" See "Not Just Another Elvis Impersonator," *New York Times*, June 16, 1991.
58. Freiburg and Freiburg, *Nuts!*, p. 60.
59. Stephen D. Solomon, "A Sharp Increase in Cabin Pressure," *New York Times*, November 28, 1993.
60. Solomon, "The Bully of the Skies Cries Uncle," p. 37.
61. Welles, "Captain Marvel."
62. See Stephen D. Solomon, "Turbulence," review of Thomas Petzinger Jr., *Hard Landing: The Epic Contest for Power and Profits That Plunged the Airlines into Chaos* (New York: Times Business Books, 1995), *New York Times*, January 7, 1996.
63. Solomon, "The Bully of the Skies," pp. 12–13.
64. Jonathan Dahl, "After the Mergers: Air Fares Rise, But Era of Bargain Rates Isn't Over," *Wall Street Journal*, February 2, 1987.
65. Ralph Blumenthal, "Southwest Air Is Facing Challenge of Its Own Making," *New York Times*, March 17, 1986 and Dennis Fulton, "Southwest Airlines Feeling Heat of Battle," *Dallas Morning News*, February 11, 1986.
66. "In 1987... [America West] had a $45 million loss as it bought new aircraft, launched domestic routes and fought a bitter fare war with Southwest Airlines." America West reacted to these losses by lowering its costs—which were second only to those of Southwest. Obviously frustrated, America West's Chairman Ed Beauvais made this remark in 1991 about Southwest: "If we could figure out how to take their head off tomorrow, we would." Bridget O'Brien, "America West Throttles Back on Growth—Financial Worries Force Carrier to Conserve Cash," *Wall Street Journal*, April 29, 1991.
67. In 1991, one industry observer wrote: "Taking a dig at America West's ardor to attain 'critical mass'—and at its 1990 net loss of $75 million—Kelleher says: Mass certainly

doesn't mean profitability. Our critical mass is our presence on each route. On our 75 top routes, we have more than 50% of the traffic, while the next carrier has 10%." Danna K. Henderson, "Southwest Luvs Passengers, Employees, Profits," *Air Transport World*, July 1991. Philip Baggaley, senior vice president of S & P CreditWatch, commented on America West in 1991: "This is a company that if they let it just operate, they do well, but their eyes are always bigger than their pocketbooks." O'Brien, "America West Throttles Back on Growth." See also Edwin McDowell, "America West Airlines Shrinks to Try to Survive," *New York Times*, August 21, 1992.

68. Muse, *Southwest Passage*, p. 223.
69. Christopher Carey, "Southwest Airlines Sets Its own Rules," *St. Louis Post-Dispatch*, December 25, 1988; Jim Barlow, "For Southwest, Sky's the Limit," *Houston Chronicle*, March 22, 1987; and Dennis Fulton, "New Muse Turning Corner, Kelleher Says," *Dallas Morning News*, October 13, 1985.
70. Kelleher even seemed to refer to the employees of Muse Air with some disdain: "Southwest pilots are paid substantially more than Muse Air Pilots.... That's understandable. Southwest pilots work for a profitable company." Fulton, "New Muse Turning Corner," *Dallas Morning News*, October 13, 1985.
71. Carey, "Southwest Airlines Sets Its own Rules." Another journalist notes: "Continental Airlines Inc. blitzed [Kelleher] with cheaper flights spaced 15 minutes before and after Muse's. By 1987, Muse was draining $2 million a month from Southwest." Kelly, "Southwest Airlines" p. 53.
72. "Southwest Air Unit Cuts All Coach Fares to Maximum of $79," *Wall Street Journal*, September 17, 1986 and Francis C. Brown III, "Southwest Air to Shut Down TranStar Aug. 9," *Wall Street Journal*, July 30, 1987.
73. The *Wall Street Journal* reported: "Mr. Kelleher said Southwest made the decision to close the unit when TranStar's losses continued into the peak summer months. 'It didn't appear that TranStar could be retuned to profitability,' Mr. Kelleher said." Brown III, "Southwest Air to Shut Down TranStar Aug. 9."
74. "One-on-One with Herb Kelleher," *Dallas Magazine*, March 1, 1986 and Dennis Fulton, "Southwest Airlines Feeling Heat of Battle," *Dallas Morning News*, February 11, 1986.
75. "The Chairman of the Board Looks Back," *Fortune*, May 28, 2001.
76. Southwest Airlines, Inc., *Southwest Airlines Annual Report for 1994*.
77. Kelly, "Southwest Airlines."
78. When Southwest first entered the California market in 1982, one journalist noted Southwest's impressive operating margin of 16 percent; as a result, "other carriers with higher costs may find themselves in a bind trying to compete." Bill Sing, "Southwest Air Heads for L.A.: Low-Cost, Low-Fare Line Bids for Major Market Share," *Los Angeles Times*, September 16, 1982. In the 3 years following deregulation, fuel costs increased an average of 66 cents, which caused financial difficulties for much of Southwest's competition. See Karr, "Airline Deregulation after Braniff's Fall," *Wall Street Journal*, June 14, 1982.
79. Danna K. Henderson, "Momentous Events Muddle the Future," *Air Transport World*, January 1, 1991.
80. In contrast, American Airlines lost $39 million, Delta $154 million, and USAir $454 million in 1990. See "Southwest Comes Out Ahead in '90," *Austin American-Statesman*, January 29, 1991.
81. Southwest Airlines, Inc., *Southwest Airlines Annual Report for 2006*.
82. Chris Kelley, "Texas Rail Hearings Likely to be Heated: Stakes High for Train, Plane Industries," *Dallas Morning News*, March 25, 1991 and "Trains on the Flight Path," *Independent*, July 30, 1995.

83. Rogelio Oliva and Jody Hoffer Gittell, "Southwest Airlines in Baltimore," HBS Case No. 602–156 (Boston: Harvard Business School Publishing, 2002), p. 3.
84. "United Fights Southwest for California Market," *Journal of Commerce*, October 14, 1994.
85. Edwin McDowell, "United Fights Back with Western Shuttle," *New York Times*, October 25, 1996.
86. Wendy Zellner, "Dogfight over California," *Business Week*, August 15, 1994.
87. Susan Chandler, "'Sky' Magary Picks a Dogfight," *Business Week*, September 19, 1994 and Scott McCartney, "Southwest Flies Circles around United's Shuttle," *Wall Street Journal*, February 20, 1996.
88. Mark Veverka, "Blood and Guts Kelleher: Southwest Chief Rallies Troops against United," *Crain's Chicago Business*, November 21, 1994.
89. McCartney, "Southwest Flies Circles."
90. Laura Goldberg, "Southwest's Approach Still Flies," *Houston Chronicle*, June 18, 2000.
91. Chris Woodyard, "There's Something Familiar about Kelleher's Successors," *USA Today*, March 20, 2001.
92. David Koenig, "Southwest Airlines Announces Lower Earnings," *Associated Press*, July 15, 2004.
93. Goldberg, "Southwest's Approach Still Flies."
94. Lori Ranson, "Southwest Ready To Grow 2006 Earnings by 15 Percent," *Aviation Daily*, January 19, 2006.

Chapter 8 Gordon Bethune's Revival of Continental Airlines

1. Portions of this chapter on Gordon Bethune and Continental Airlines have been excerpted from Nitin Nohria, Anthony J. Mayo, and Mark Benson, "Gordon Bethune at Continental Airlines," HBS Case No. 406–073 (Boston: Harvard Business School Publishing, 2006).
2. Continental Airlines, Inc., *Continental Annual Reports for 1993–1994*.
3. Greg Brenneman, "Right Away and All at Once: How We Saved Continental," *Harvard Business Review*, September–October 1998, p. 4.
4. Ibid., p. 12.
5. Rosabeth Moss Kanter, *Confidence: How Winning and Losing Streaks Begin and End* (New York: Crown Business, 2005), p. 230.
6. Delta's purchase of the remainder of Pan Am's operations was the largest acquisition of flights in airline history. "Delta Timeline," The Delta Heritage Museum Web site, http://www.deltamuseum.org/M_Education_DeltaHistory_Facts_Timeline.htm (accessed June 23, 2008).
7. James Cook, "Lorenzo the Presumptuous," *Forbes*, October 1978, pp. 115–117.
8. Sources for the sidebar on Robert Six include Robert J. Serling, *Maverick* (New York: Doubleday, 1974); R. E. G. Davies, *Continental Airlines: The First Fifty Years, 1934–1984* (The Woodlands, TX: Pioneer Publications, 1984); and Robert F. Six, *Continental Airlines: A Story of Growth* (New York: Newcomen Society, 1959).
9. The merger was officially approved by shareholders and was structured in the following way: Texas Air Corporation, also called Texas Air, was incorporated in October 1982, as the parent company for Continental Airlines Corporation (CAC), also organized in October 1982. CAC had two subsidiaries—Continental Air Lines, Inc. (CAL) and Texas International Airlines Inc. (TXI).
10. Kanter, *Confidence*, p. 231.

11. Lorenzo at November 12, 1984 Question & Answer Session with participants in Harvard Business School's Advanced Management Program.
12. Continental Airlines, Inc., *Continental Annual Reports, 1978–1980, and Continental 10K Filings, 1978–1985.*
13. Thomas Petzinger Jr., *Hard Landing: The Epic Contest for Power and Profits That Plunged the Airlines into Chaos* (New York: Times Business Books, 1995), p. 335.
14. Eric Weiner, "Lorenzo, Head of Continental Air, Quits Industry in $30 Million Deal," *New York Times*, August 10, 1990 and Floyd Norris, "A Good Deal? Yes, for Lorenzo," *New York Times*, August 10, 1990.
15. Wendy Zellner, "This Is Captain Ferguson, Please Hang on to Your Hats," *Business Week*, May 25, 1994, p. 54.
16. Adam Bryant, "Continental is Dropping 'Lite' Service," *New York Times*, April 14, 1995.
17. Ibid.
18. "Continental Taps Boeing Executive to Be Chief Executive Officer," *Aviation Daily*, January 31, 1994.
19. Brenneman, "Right Away and All at Once," p. 4.
20. Gordon Bethune, *From Worst to First: Behind the Scenes of Continental's Remarkable Comeback* (New York: John J. Wiley, 1998), p. 248.
21. Ibid., p. 250.
22. Ibid., p. 251.
23. Ibid., p. 253.
24. Martin Puris, *Comeback: How Seven Straight-Shooting CEOs Turned Around Troubled Companies* (New York: Random House-Times Business, 1999), p. 25.
25. "Braniff Sees Record Profit of $15 Million in Third Quarter; Plans to Buy 22 Boeing Jets," *Wall Street Journal*, September 22, 1978, p. 6.
26. Company history information available at www.braniffinternational.com (accessed December 2005).
27. "Braniff International Earnings Declined 21% in the Second Quarter," *Wall Street Journal*, July 23, 1979, p. 8 and "Braniff Had Loss of $51.4 Million for 4th Quarter," *Wall Street Journal*, February 11, 1980, p. 14.
28. "Braniff International Earnings," p. 8.
29. Bethune, *From Worst to First*, pp. 150–151.
30. Ibid., p. 264.
31. Ibid., p. 268.
32. "Boeing 737-X Nears Go Ahead," *Aviation Week*, June 21, 1993, p. 30.
33. Bethune, *From Worst to First*, p. 17.
34. Brenneman, "Right Away and All at Once," p. 6.
35. Bethune, *From Worst to First*, pp. 22–23.
36. Ibid., pp. 22–23 and Brenneman, "Right Away and All at Once," p. 4.
37. Continental Airlines, Inc., *Continental Airlines 10-K Filing with the Securities and Exchange Commission for 1994*, pp. 3–4.
38. Bethune, *From Worst to First*, p. 23.
39. Ibid., p. 47.
40. "Continental Scraps 'Lite' Experiment, Takes Huge Loss," *Oklahoma City Journal Record*, April 14, 1995.
41. Bethune, *From Worst to First*, p. 50 and Thomas J. Neff and James M. Citrin, *Lessons from the Top: The 50 Most Successful Business Leaders in America – and What You Can Learn from Them* (New York: Currency Doubleday, 1999), p. 58.
42. Bethune, *From Worst to First*, p. 63.

43. Brenneman, "Right Away and All at Once," p. 6.
44. "Continental to Save $150 Million a Year with Further Capacity Cuts," *Airports*, January 3, 1995, p. 4 and "Continental Plans 4,000 Job Cuts as It Clips Capacity," *Aviation Daily*, January 18, 1995, p. 85.
45. Bethune, *From Worst to First*, pp. 84–85.
46. Brenneman, "Right Away and All at Once," p. 7.
47. Jonathon Burton, "An Outsider in Continental's Cockpit? That's OK, He Knows How to Fly," *Chief Executive*, December 1995, p. 26.
48. Gordon Bethune, *Continental Airlines: From Worst to First: The Remarkable Turnaround Story of an American Corporation* (New York: Newcomen Society of the United States, 1999), p. 13.
49. Brenneman, "Right Away and All at Once," p. 6.
50. Bethune, *From Worst to First*, p. 101.
51. Ibid., p. 115.
52. Ibid., p. 114.
53. Kanter, *Confidence*, p. 231.
54. Brenneman, "Right Away and All at Once," p. 7.
55. "Gordon Bethune, Chairman and Chief Executive Officer of Continental Airlines," *Aviation Week & Space Technology*, April 7, 1997.
56. *CEO Exchange: Mastering the Art of Corporate Reinvention*, WTTW-Chicago Video, 2000.
57. Bethune, *From Worst to First*, p. 105.
58. Ibid., p. 193.
59. Brenneman, "Right Away and All at Once," p. 11.
60. Bethune, *From Worst to First*, pp. 36–37.
61. Ibid., p. 141.
62. Brenneman, "Right Away and All at Once," p. 11.
63. Burton, "An Outsider in Continental's Cockpit? That's OK, He Knows How to Fly," *Chief Executive*, December 1995, p. 26.
64. Continental Airlines, Inc., *Continental Airlines Annual Report for 2000*, p. 8.
65. Ibid., p. 9.
66. Al Frank, "Continental Airlines' Gordon Bethune Faces His Biggest Challenge," *Newark Star Ledger*, October 26, 2001.
67. "Continental Expects to Avert Over 1,000 Furloughs; Severance Costs Total Over $60 Million," *PR Newswire*, September 26, 2001.
68. Ibid.
69. Ibid.
70. "Bush Signs Airline Bailout Package," *CNN*, September 23, 2001.
71. "Continental Airlines Reports Fourth Quarter and Full Year Loss," *PR Newswire*, January 15, 2003.
72. "Bethune Defiant as Continental Plugs Pensions Gap," *Airline Business*, October 2003, p. 12.
73. Bill Hensel, Jr., "A Hometown Inauguration: Changing Guard at Continental," *Houston Chronicle*, December 26, 2004, p. 1.

Epilogue

1. Almost a quarter century ago, Paul Lawrence and Davis Dyer described "mutual adaptation" as a process by which organizations both impact the nature of their

overarching industrial environment and adapt in response to evolving contextual factors. Lawrence and Dyer further contended that businesses should aim to put themselves under a moderate amount of pressure to integrate a "readaptive process" into the center of the organization. This could be achieved by monitoring the level of complexity a firm has to deal with as well as the amount of resources it has at its disposal. In their study of American industry, they identified how certain organizations created the capacity for change and adaptability. Their study of organizational adaptation cited the impact of "key acts of leadership, crucial governmental impacts, and important organizational practices and strategies." See Paul R. Lawrence and Davis Dyer, *Renewing American Industry* (New York: Free Press, 1983), pp. x–xi, 4, 9. Twenty years later, Warren G. Bennis and Robert J. Thomas cited a similar characteristic of successful and resilient business leaders that they called "adaptive capacity"—the ability to face adversity or change and to modify behavior to ensure continued success. See Warren G. Bennis and Robert J. Thomas, *Geeks & Geezers: How Era, Values, and Defining Moments Shape Leaders* (Boston: Harvard Business School Press, 2002), pp. 18–19. To sustain success through industry lifecycle stages, the leader must be cognizant of both internal and external forces and be prepared to act. We called this awareness of and ability to adapt to the context as "contextual intelligence." See Anthony J. Mayo and Nitin Nohria, *In Their Time: The Greatest Business Leaders of the 20th Century* (Boston: Harvard Business School Press, 2005), p. xv.

2. Anita McGahan, "How Industries Change," *Harvard Business Review*, October 2004. McGahan notes that "Over time, the industry experiences a *shakeout*, usually because a specific business model achieves greater legitimacy than any other."

Bibliography

Ahouse, Jeremy C. "The Tragedy of *a priori* Selectionism: Dennett and Gould on Adaptationism." *Biology and Philosophy* 13 (1998): 359–391.

Alchian, Armen A. "Uncertainty, Evolution, and Economic Theory." *Journal of Political Economy* 53 no. 3 (1950): 211–221.

Baldwin, Carliss Y., and Kim B. Clark. "Managing in an Age of Modularity." *Harvard Business Review* (September 1997): 84–93.

Bennis, Warren, and Burt Nanus. *Leaders: The Strategies for Taking Charge.* New York: Harper & Row, 1985.

Chandler Jr., Alfred D. *Strategy and Structure: Chapters in the History of Industrial Enterprise.* Cambridge, MA: MIT Press, 1961.

Chandler Jr., Alfred D. with Takashu Hikino. *Scale and Scope: The Dynamics of Industrial Capitalism.* Cambridge, MA: Belknap Press, 1990.

Christensen, Clayton M., and Joseph L. Bower. "Customer Power, Strategic Investment, and the Failure of Leading Firms." *Strategic Management Journal* 17 no. 3 (March 1996): 197–218.

Christensen, Clayton M., Scott D. Anthony, and Erik A. Roth. *Seeing What's Next: Using the Theories of Innovation to Predict Industry Change.* Boston: Harvard Business School Press, 2004.

DiMaggio, Paul J., and Walter W. Powell. "The Iron Cage Revisited: Institutional Isomorphism and Collective Rationality in Organizational Fields." *American Sociological Review* 48 no. 2 (April 1983): 147–160.

Dosi, Giovanni, Richard R. Nelson, and Sidney G. Winter. *The Nature and Dynamics of Organizational Capabilities.* Oxford: Oxford University Press, 2000.

Eaton, B. Curtis. "Review of an Evolutionary Theory of Economic Change." *Canadian Journal of Economics* 17 no. 4 (November 1984): 868–871.

Gardner, Howard with Emma Laskin. *Leading Minds: An Anatomy of Leadership.* New York: Basic Books, 1995.

Gould, Steven Jay, and Niles Eldredge. "Punctuated Equilibria: The Tempo and Mode of Evolution Reconsidered." *Paleobiology* 3 no. 2 (Spring 1977): 115–151.

Greiner, Larry E. "Evolution and Revolution as Organizations Grow." *Harvard Business Review* 76 no. 3 (May–June 1998): 55–67.

Hannan, Michael T., and John Freeman. "The Population Ecology of Organizations." *American Journal of Sociology* 82 no. 5 (March 1977): 929–964.

Harrison, J. Richard, and Glenn R. Carroll. *Culture and Demography in Organizations.* Princeton, NJ: Princeton University Press, 2006.

Hayward, Matthew L. A., Dean A. Shepherd, and Dale Griffin. "A Hubris Theory of Entrepreneurship." *Management Science* 52 no. 2 (February 2006): 160–172.

Henderson, Rebecca M., and Kim B. Clark. "Architectural Innovation: The Reconfiguration of Existing Product Technologies and the Failure of Established Firms." *Administrative Science Quarterly* 35 no. 1 (March 1990): 9–30.

Hiller, Nathan J., and Donald C. Hambrick. "Conceptualizing Executive Hubris: The Role of (Hyper-) Core Self-Evaluations in Strategic Decision-Making." *Strategic Management Journal* 26 (2005): 297–319.

Jawahar, I. M., and Gary L. McLaughlin. "Toward a Descriptive Stakeholder Theory: An Organizational Life Cycle Approach." *Academy of Management Review* 26 no. 3 (July 2001): 397–414.

Kotter, John P. *A Force for Change: How Leadership Differs from Management.* New York: Free Press, 1990.

Lawrence, Paul R., and Davis Dyer. *Renewing American Industry.* New York: Free Press, 1983.

Levinthal, Daniel A. "The Slow Pace of Rapid Technological Change: Gradualism and Punctuation in Technological Change." *Industrial and Corporate Change* 7 no. 2 (1998): 217–247.

Lynall, Michael D., Brian R. Golden, and Amy J. Hillman. "Board Composition from Adolescence to Maturity: A Multi-Theoretic View." *Academy of Management Review* 28 no. 3 (July 2003): 416–431.

McGahan, Anita M. "How Industries Change." *Harvard Business Review* 80 (October 2004): 86–94.

Mayo, Anthony J., and Nitin Nohria. *In Their Time: The Greatest Business Leaders of the Twentieth Century.* Boston: Harvard Business School Press, 2005.

Meyer, John W., and Brian Rowan. "Institutionalized Organizations: Formal Structure as Myth and Ceremony." *American Journal of Sociology* 83 no. 2 (September 1977): 340–363.

Mirowski, Philip. "An Evolutionary Theory of Economic Change: A Review Article." *Journal of Economic Issues* 17 no. 3 (September 1983): 757–768.

Nakamura, Leonard I. "Economics and the New Economy: The Invisible Hand Meets Creative Destruction." *Business Review—Federal Reserve Bank of Philadelphia* (July– August 2000): 15–31.

Nelson, Richard R., and Sidney G. Winter. "The Schumpeterian Tradeoff Revisited." *American Economic Review* 72 no. 1 (1982): 114–132.

———. *An Evolutionary Theory of Economic Change.* Cambridge, MA: Harvard University Press, 1982.

Quinn, Robert E., and Kim Cameron. "Organizational Life Cycles and Shifting Criteria of Effectiveness: Some Preliminary Evidence." *Management Science* 29 no. 1 (January 1983): 33–51.

Schumpeter, Joseph. "The Instability of Capitalism," *Economic Journal* 38 no. 151 (September 1928): 361–386.

———. *The Theory of Economic Development: An Inquiry into Profits, Capital, Credit, Interest, and the Business Cycle.* Cambridge, MA: Harvard University Press, 1934.

Smith, Ken G., Terence R. Mitchell, and Charles E. Summer. "Top Level Management Priorities in Different Stages of the Organizational Life Cycle." *Academy of Management Journal* 28 no. 4 (December 1985): 799–820.

Stinchcombe, Arthur L. "The Sociology of Organization and the Theory of the Firm." *Pacific Sociological Review* 3 no. 2 (Autumn 1960): 75–82.

Tushman, Michael L., William H. Newman, and Elaine Romanelli. "Convergence and Upheaval: Managing the Unsteady Pace of Organizational Evolution." *California Management Review* 29 no. 1 (Fall 1986): 29–45.

Veblen, Thorstein. "Why Is Economics Not an Evolutionary Science?" *Quarterly Journal of Economics* 12 no. 4 (July 1898): 373–397.

Index